ARCHITECTURAL SURFACES
DETAILS FOR ARTISTS, ARCHITECTS, AND DESIGNERS

建筑表皮
设计师与艺术家的设计元素

（美）朱蒂・A．朱瑞克 编
Judy A. Juracek

刘冠楠 译

大连理工大学出版社

Architectural Surfaces: Details for Artists, Architects, and Designers
by Judy A. Juracek
©2005 by Judy A. Juracek
The translation of Architectural Surfaces: Details for Artists, Architects, and
Designers is published by arrangement with W.W.Norton & Company, Inc
©大连理工大学出版社2010
著作权合同登记06-2007年第180号

图书在版编目(CIP)数据

建筑表皮 : 设计师与艺术家的设计元素 : 汉英对照
/ (美) 朱瑞克 (Juracek,J.A.) 编 ; 刘冠楠译. 一 大
连 : 大连理工大学出版社, 2010.7
    ISBN 978-7-5611-5564-6

    I. ①建… Ⅱ. ①朱… ②刘… Ⅲ. ①建筑设计—汉
、英 Ⅳ. ①TU2

    中国版本图书馆CIP数据核字（2010）第102975号

出版发行：大连理工大学出版社
            (地址：大连市软件园路80号 邮编：116023)
印   刷：利丰雅高印刷（深圳）有限公司
幅面尺寸：216mm×280mm
印   张：21
插   页：4
出版时间：2010年7月第1版
印刷时间：2010年7月第1次印刷
责任编辑：袁 斌
责任校对：马 帅
封面设计：温广强

ISBN 978-7-5611-5564-6
定   价：228.00元

电 话：0411-84708842
传 真：0411-84701466
邮 购：0411-84703636
E-mail: designbook@yahoo.cn
URL: http://www.dutp.cn

如有质量问题请联系出版中心：（0411）84709043 84709246

ARCHITECTURAL SURFACES

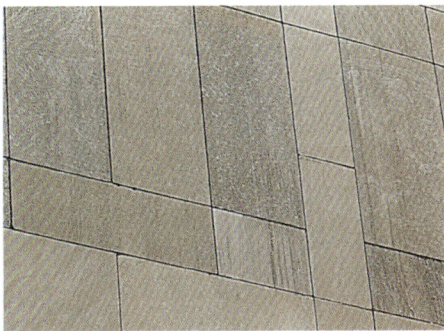

目录

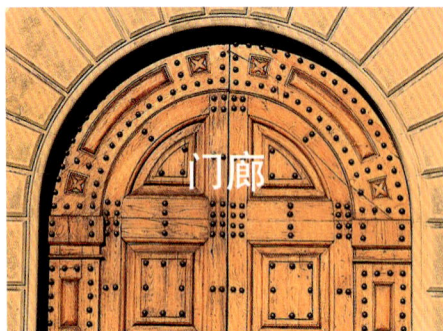

前言
彼得·派诺耶

随着本书的出版，朱蒂·朱瑞克（Judy Juracek）完成了她对自然界和人造世界形象的一系列收藏的第四卷。这部书提供了清晰明了的物体图像的照片。这本书的副标题为"设计师与艺术家的设计元素"。事实上，本书不仅是一本研究工具书，它还捕捉到了大量不同类别的建筑特点，提供给人以启发性的照片。作为一本给建筑师、室内设计师、景观设计师和相关专业人士的工具书，这本书在很多层面上都很有价值。

植物和动物的目录具有清晰的组织结构，因为每一个种类都有其特定的区别于其他种类的特征。树、鸟和兽类的指南几乎每年都会出现，但是将建筑元素编成目录则无疑是更有难度的，它要求作者在以个人判断代替分类学来进行分类的领域里冒险。在《1780年以来的美国建筑》一书中，马库斯·维峰（Marcus Wiffen）描述了四十种建筑风格，但同时承认"每一种风格都有一种几乎无穷的混合能力"。本书允许每一个部分的独立特征突现出来，同时也表明它与建筑整体之间的关系。

朱瑞克设法给这个特别的图像目录书籍编写一个有用的顺序，但并没有破坏美感。书中的分类不仅能充当内容上优雅的向导，且看起来从未妨碍视觉浏览的乐趣。在这样的排版方式下，我们的目光从日本石柱移到爱奥尼亚柱基，二者的共性得到增强。通过储存抽象的建筑知识和背景原料，更是依赖艺术家的眼光，她获得了这种平衡。

建筑师和设计师花费很长时间为他们自己的图书库搜集重要的图片。巴黎Ecole des Beux-Arts——1850～1950年间世界上最伟大的建筑学校——在校学生可以进行大规模的旅行：向南穿过意大利去观赏、测量和描绘伟大的历史遗址，这些地方通常是当地并不著名的建筑结构。他们的素描本成为未来证明的工具和很多伟大实践的基础。这些图像也用于使其他的、更学术的资源更易接受且更真实。当人们观赏这些图片并关注于它们的结构、色彩和自然建筑的物质性时，素描中的伟大的文艺复兴建筑在保罗·莱特茹里（Paul Letarouilly）的《现代罗马大厦》中重生。

一旦能够获得建筑照片，建筑师和设计师们就实现了他们重现记忆影像的价值。查尔斯·普拉特（Charles Platt）在1890年成为建筑师的艺术家，虽然没能拥有在Ecole des Beux-Arts求学的机会，但是他同他的哥哥一起游遍欧洲，收集了大量的图片（一些是自己拍摄的，另一些是购买的），这些图片成为了他办公室图书馆的精髓。这些图片按主题分类被装订成书卷——喷泉、拱顶、天花板等——而不是按照历史或学术顺序。这些书虽然已经破旧褪色，但作为设计师早期尝试捕捉事物样貌的证据保留到现在。但是即使是普拉特的收集也只占朱瑞克在本书中的收录图片的一小部分。

这本书中每一张照片都代表着一个主题的自然特征：如果是石头，我们可以看到条纹、色彩、纹理和形状。这些基本属性是我们在任何书目中都能看到的，但在这里我们看到的更多。朱瑞克的镜头表现出使图像更富有生气的性质：时间和地点、强度和太阳视角、光照、空气等。

这本书最令人惊讶的成就是除少数几张照片、之外，大量的照片全是由作者拍摄的。她摄影师般的眼光在她艺术家般的作品中受到鼓舞，更具透视效果。她连贯的观点使得这一系列图像有一种独特的风格，豪不夸张地把它们作为一个作品的整体联系在一起。

当互联网作为设计师的工具变得更受欢迎时，朱瑞克的书为印刷书籍的美好和实用性注入令人耳目一新的力量。这一卷对于设计界的设计资料藏书是一个受欢迎的增补。带着任务翻看这本书——例如，要找一种特别的木质镶板，同时也会发现搜索材料之外的部分。这就是本书带给读者的意外惊喜。

最后，任何花时间看这本书的设计师都会发现另一个明显的特征：有这些图片在手，能够激励人们更近距离观察世界，并像朱蒂·朱瑞克所做的那样注意到我们四周里建筑表皮的本质特征。

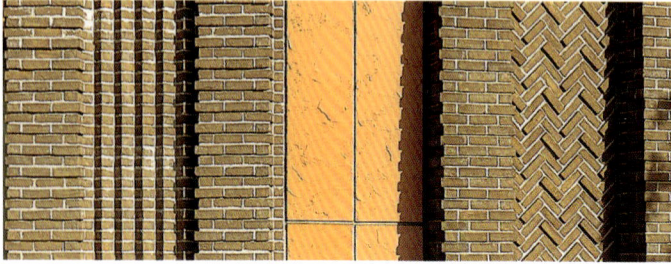

鸣谢

像本书这一类型的书要求有广泛的资源，如果没有这里提到的这些专家意见、建议和慷慨的帮助，那么，这本书将不可能存在。除了要表达我对他们的贡献最深切的感激之外，我还必须说，同这些人的相遇并合作是非常愉快的经历。

在这次工作开始时，我的编辑南茜·格林建议我请彼得·派诺耶作建筑顾问，这也是我一直所感激的建议。他为本书的结构和图片的编辑与组织提供了一个建筑师的观点。除了谨慎的技术上的建议，他对该项目的热忱和对相关事物的领悟深入到了工作中，对本书的完成做出了巨大的贡献。

格·葛尼（Guy Gurney）帮助我为这四本书照相，他不仅在技术设备和影像方面给了我大量的建议，而且在灯光和拍摄地点方面也给予了我很大帮助。此外，他还向这些计划开放了他的照片存档。如果没有他，表皮系列丛书中的许多影像都无法寻得其最佳方式。

其他对这本书的图片做出贡献的是麦迪斯·巴查特（Medith Barchat）、杜安·朗根瓦尔特（Duane Langenwalter）和彼得·米勒(Peter Miller)，他们都是对寻找建筑影像有特殊爱好的优秀摄影师。

奥斯汀建筑公司的朱迪·璞克顿（Judy Procton）、派特森（Patterson），蒂斯顿（Disstoon）能够理解我正在寻找的细节，寻找他们公司存档中的细节实例，并慷慨地安排到书的相应部分中。约翰·温菲尔德（John Winfield）也帮助我为照片安排位置。并且，纽波特保护协会（the Newport Preservation Society）和洛克伍德·马修士庄园博物馆（the Lock wood–Mathews Mansion）允许我使用他们的收藏品中独特的建筑。

简和威廉·英格拉姆（Jean and William Ingram）组织了我对巴厘岛的一些特殊建筑结构的拍摄，并获得Begawan Giri度假酒店、The Tuga 度假酒店和Puri Lumbung度假酒店的拍摄许可。我在巴厘岛的导游，梅德·姆拉(Made Murra)在寻找建筑细部上拥有真正的诀窍。在东京，摄影师Kaoru Soehata和Tomoko Shiozawa不仅知道我想要拍什么，而且知道拍摄的最佳时间；南希·凯特苏拉（Nancy Katsura）为我设计了一次卓绝的城市建筑之旅；Yukiko Tanisho和Sumiko Nakamura让我分享了他们对京都建筑的理解；Toshitaka Kitayma和Naoko Wada在这本书中公开了Agon Shu宗教组织。在洛杉矶，伊丽莎白·纽厄尔（Elisabeth Newell）提出她建筑方面的观点和知识；凯伦和弗兰克·阿米蒂奇还教给了我另一个丰富的摄影路线。

威廉姆·怀斯特（William West）和乔治·达伯瑞德（George de Brigard）对本书进行文字说明，我被他们的知识和坚定不移的信念所震撼。砖石工业组织的布莱恩·崔姆百尔（Brian Trimbel）同O&G工业的安迪·巴诺（Andy Banores）一样，提供了专业的文字说明信息。还要感谢戴安·瑞奇，我（Diane Rich）分享了她在阅读说明方面的技术。

这本书同样是出自于其他表皮系列丛书的工作组，而我永远无法完全的表达出我对与他们技术和贡献的感激。南茜·格林对于她的支持和书的信仰是忠诚的，里阿·格雷厄姆（Leeann Graham）在出版上继续给予了悉心的指导，鲍勃·埃尔伍德（Bob Elwood）的索引对于这一系列是一个特别的贡献，而Gilda Hannah也再次帮助编辑了图片和版面设计，提高了影像的质量，促使读者想要继续阅读下一页。

我同样感激巴巴拉·勃朗（Barbara Braun）的协助和伊丽莎白·沃尔（Elizabeth Woll）设计护封。

简介

一个主题公园的设计者曾经告诉我，设计最难的部分就是把棱角处完美地结合到一起。她补充到，任何一个有天赋的人都能做出一个可使用的设计品或画出一副精美的建筑正视图，然而设计师的技巧就在于他如何处理细节。一块块的砖是如何拼凑成拱形的？两面石头墙是如何完美地相联在一起？上楣包裹的建筑拐角处模型的侧面是什么样的？能够独特而典雅地解决这些问题是一个好的设计的标志；解决方案通常是一些结构性的装饰细节，它们代表一定的历史时期，并形成设计师独特的个人风格。

所有的艺术家，包括插图画家、漫画家和景观设计师，都会探究建筑细节，就好像他们复制或再现了情景和环境。使用合适风格的窗户，适宜的地板和磨光的墙壁通常标志着艺术家作品的真实性。现在，对于一部成功的设计来说精确性是特别重要的，因为我们曾经经历过视觉信息大爆炸，也就是说很少有人没被大量的图像影响过。因此，各年龄段的观众都是非常老练的。他们可能是通过耳濡目染获得这方面的知识，他们所认为存在的建筑都是从影片、主题公园或是电脑游戏环境下构造的。

本书是一部建筑图集，它关注的焦点在于某一个特定的细节或建筑元素上。在大部分情况下，图片组织从总体上说是很明显的："窗户"是一个章节，"天花板和屋顶"是另外一个章节。然而，到涉及到墙面时，很明显这样的图像排版就略显笨拙，不实用。因此这部分图片排版划分成两个部分——"墙面"和"外立面"。第一部分包括由石头、木头和砖等建筑材料制成的物体。第二部分是由如玻璃、金属隔墙和建筑用的赤土陶器等材料应用于建筑中构成的。这种分类方法也具有一定的任意性，例如，石质的胶合板

（"墙面"）在技术上来说也和其他的隔墙同属一类（"外立面"）。但是，砂石被看做是一种结构性材料，在石质胶合板的粘合过程的细节中，它又是跟固体的石质建筑是一样的。

如果把图片按视觉相似主题划分为特定的类别进行归档，图示研究使用起来就会比较容易。带着这种思想，各章节会被进一步分为几个小部分，分别关注同一主题的不同部分。例如，"门廊"被分为镶板门、板条门、喷漆门、金属门、顶窗、边窗、入口、大门和门饰。这些分类不是为了解释"门廊"的各个不同的部分，而是为了将门和出口的各个不同部分分割出来以便展示该种类内的各种风格。

在这种结构下，目录是非常有用的搜索引擎，它能够帮助使用者找到确切的建筑细部。在说明中的主题关键字都在索引中出现过。建筑因素涵盖了各种类，而且并不是一个主题的所有范例都限制在一个章节中。例如，"天花板和屋顶"那一章的照片中，一个有趣的帕拉斯式的窗户展示了一种独特的屋顶窗。然而，窗户的风格是在说明中命名的，在目录中也会找到对应的参照。

说明部分是在专家的帮助下编写的，并从各种资源获得多方面的信息。由于大部分的识别都仅仅来源于照片，不可避免会有一些纰漏。

书里的图片都是作者的，或是图片鸣谢中所列举的个人或组织的私有财产，未经同意，不得以任何形式用于商业行为。

参考书目

American Architecture: An Illustrated Encyclopedia. Cyril M. Harris. New York: W. W. Norton & Company, 1998.

American House Styles: A Concise Guide. John Milnes Baker. New York: W. W. Norton & Company, 1994.

The Architecture of Japan. Arthur Drexler. New York: The Museum of Modern Art, 1955

The Architecture Traveler: A Guide to 263 Key American Buildings. Sydney LeBlanc. New York: W. W. Norton & Company, 2005.

ArchitectureWeek: The New Magazine of Design and Building. www.ArchitectureWeek.com.

Architecture Woodwork Quality Standards, 8th ed. Reston, VA: the Architectural Woodwork Institute, 2003.

Building Construction Illustrated, Francis D. K. Ching. New York: Van Nostrand Reinhold Company, 1975.

The Classical Orders of Architecture. Robert Chitham. New York: Rizzoli, 1985.

Construction Glossary. J. Stewart Stein. New York: John Wiley and Sons, 1980.

Dictionary of Architecture. James Stevens Curl. Oxford: Oxford University Press, 1999.

Dictionary of Architecture and Construction, 2nd ed. Cyril Harris. New York: McGraw-Hill, 1993

The Encyclopedia of Decorative Arts: 1890-1940. Philippe Garner. New York: Van Nostrand Reinhold Company, 1978

The Elements of Style, rev. ed. Stephen Calloway, Elizabeth Cromley, editors. New York: Simon & Schuster, 1996.

Glossary of Stone Industry Terms. Building Stone Institute. Itasca, IL: Building Stone Institute, 1997.

A Guide to Paving. AJ McCormack and Son. http://www.pavingexpert.com/home.htm.

A History of Architecture. Spiro Kostof. New York: Oxford University Press, 1985.

Illustrated Dictionary of Building Materials and Techniques. Paul Bianchina. Blue Ridge Summit, PA: Tab Books, 1986.

Katachi: Classical Japanese Design. Takeji Iwamiya and Kazuya Takaoka. San Francisco: Chronicle Books, 1999.

The Masonry Glossary. International Masonry Institute, M. Patricia Cronin, editor. Boston: CBI Publishing Company, 1981.

Pictorial Encyclopedia of Historical Architectural Plans, Details and Elements. John Theodore Haneman. New York: Dover Publications, 1984.

Pictorial Glossary: Architecture. The Heritage Education Network (THEN). http://histpres.mtsu.edu/then/Architecture/index.html.

Tile Glossary. Interceramic. http://www.interceramicusa.com/us/en/ products/ti/tile glossary.asp.

A Visual Dictionary of Architecture. Frank Ching. New York: Van Nostrand Reinhold Company, 1995.

The Woodworker s Reference Guide and Sourcebook. John L. Feirer. New York: Charles Scribner s Sons, 1983.

图片鸣谢

Austin, Patterson, Disston
376 Pequot Avenue
Southport, CT 06890
203-255-4031
  and
4 Midland Street
Quogue, NY 11959
631-653-1481
www.apdarchitects.com

Meredith Barchat
For picture information: www.stocksurfaces.com

Roger Bartels Architects
27 Elizabeth Street
Norwalk, CT 06854
203-838-5517
www.rogerbartelsarchitects.com

Dobyan & Dobyan Builders
Fairfield, CT 06824
203-395-8553.

Guy Gurney
P.O. Box 1227
Darien, CT 06820
203-656-6652
www.guygurney.com

Duane Langenwalter
477 Main Street, Suite 204
Monroe, CT 06468-1139
www.outofthinair.com

The Lockwood-Mathews Mansion Museum
295 West Avenue
Norwalk, CT 06850
203 838-9799

Peter Miller
For picture information: www.stocksurfaces.com

The Preservation Society of Newport County
424 Bellevue Avenue
Newport, RI 02840
401-847-1000

墙面

摄影：Austin, Patterson, Disston

WA-1 喷涂木制框架上带有完美方形边缘的室内樱桃贴面嵌板
Interior applied cherry-veneered panels with square finished edges on painted wood frame

WA-2 风化杉木屏风。顶部：V形垂直木板；底部：人字形图案
Weathered cedar-paneled entry screen. Top: notched vertical boards; bottom: herringbone pattern

WA-3 风化杉木镶板，垂直舌形槽
Weathered cedar siding, vertical tongue-and-groove

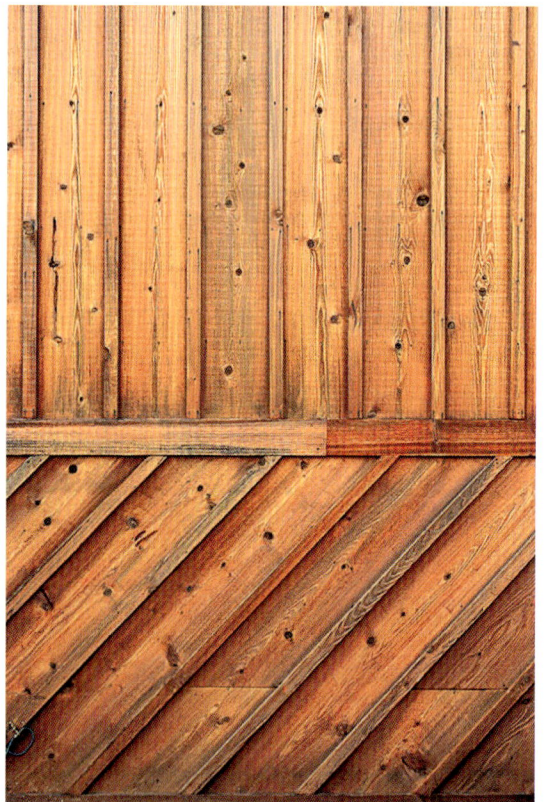

WA-4 彩色封蜡木板和板条镶板
Stained and sealed board-and-batten siding

木制墙面和镶板

WA-5 松木原木房屋结构,角落细部
Pine log house construction, corner detail

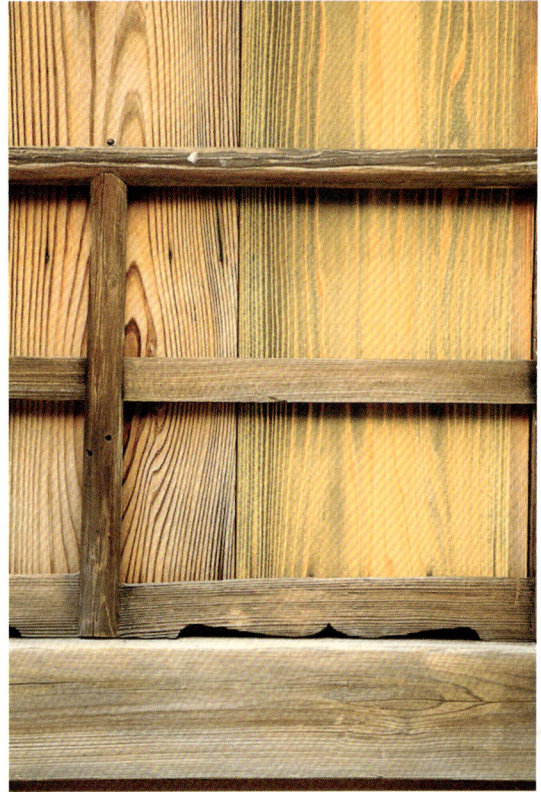

WA-6 带有装饰板条的弦切垂直厚木板
Plain-sawn vertical planks with decorative battens

WA-7 多层杉木的齐平弦切板墙,配有水平分割板条
Knotty-cedar flush, vertical plain-sawn board wall, with horizontal batten divider

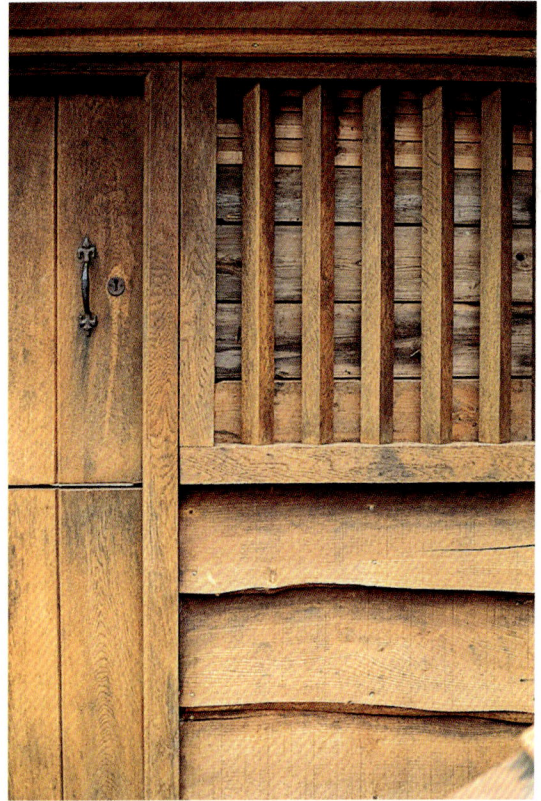

WA-8 板条门(左),百叶窗在对角分裂的老式木栅后(右上);月亮形切割的雨幕风檐板底部(右下)在附属建筑物上
Batten door (left); shutter behind diagonally split square-stock wood bars (upper right); elm waney-edged clapboard bottom (lower right) on an outbuilding

木制墙面和镶板

WA-9 方形木框架形成V字形切口拐角连接，喷漆杉木叠嵌在混凝土基座上
Notched corner joint of square timber frame with painted cedar shakes on concrete foundation

WA-10 木制框架结构，填入装有陶瓷通风口的弦切式厚杉木板
Timber-frame construction with infill of plain-sawn cedar planks with ceramic ventilation opening

WA-11 喷漆风化木镶板，左：木板和板条；右：隔板
Weathered painted wood siding. Left: board-and-batten; right: clapboard

WA-12 编制的竹条系在竹子支柱和横杆上，花园栅栏
Woven bamboo twigs tied to bamboo post and rails, garden fence

WA-13 编织的竹条墙，热带棚屋
Woven split-bamboo wall, tropical hut

摄影: Guy Gurney

WA-14 弦切垂直木板栅栏
Plain-sawn vertical board fence

木制墙面和镶板

WA-15 山墙上装有新的杉木的垂直木镶板，住宅车库
Vertical board siding with new cedar shakes on gable, residential garage

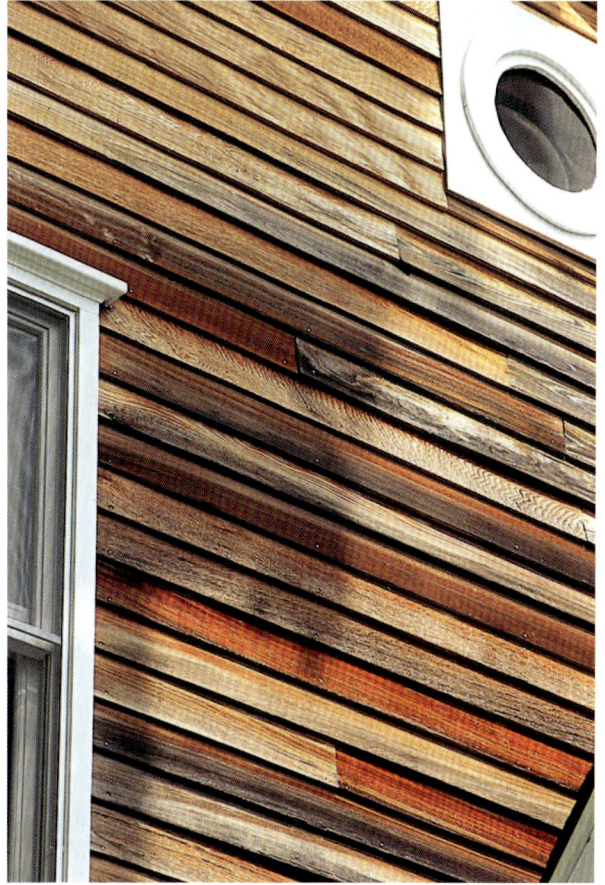

WA-16 未喷漆的墙板和窗户外部细部
Detail of unpainted clapboard and window casing

WA-17 石头根基上的箱式弦切厚木板入口墙
Low entry wall of boxed plain-sawn planks on stone base

木制墙面和镶板

WA-18 拼花设计的门扇
Entry screen with parquet design

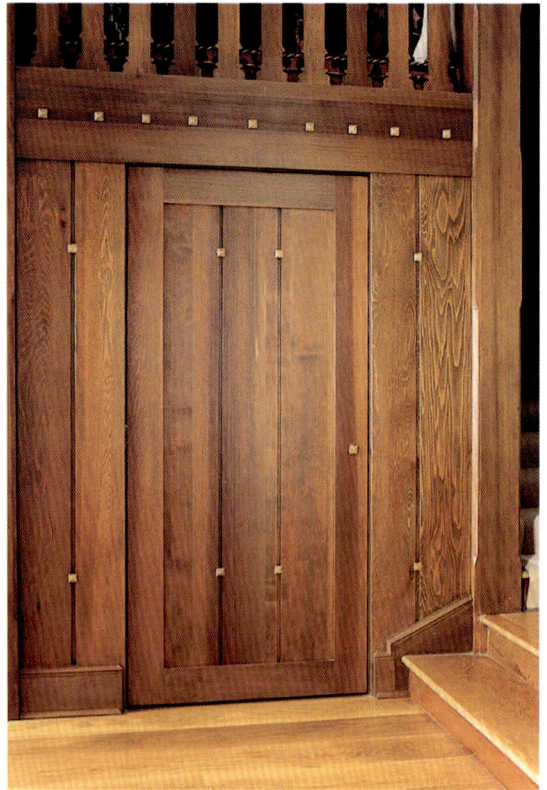

WA-19 框架式漆油杉木垂直镶板，室内门厅
Framed oiled cedar vertical board paneling, interior hallway

WA-20 框架式竹制花园围墙
Framed bamboo garden wall

WA-21 装有斜接角的古典栗色室内嵌板和突出嵌板
Antique chestnut interior paneling with mitered corners and raised panels

摄影: Austin, Patterson, Disston

木制墙面和镶板

WA-22 饰有山墙装饰和裸露的垂直板条覆层的喷漆山墙隔板
Painted clapboard gable-end wall with gable ornament and exposed vertical board cladding

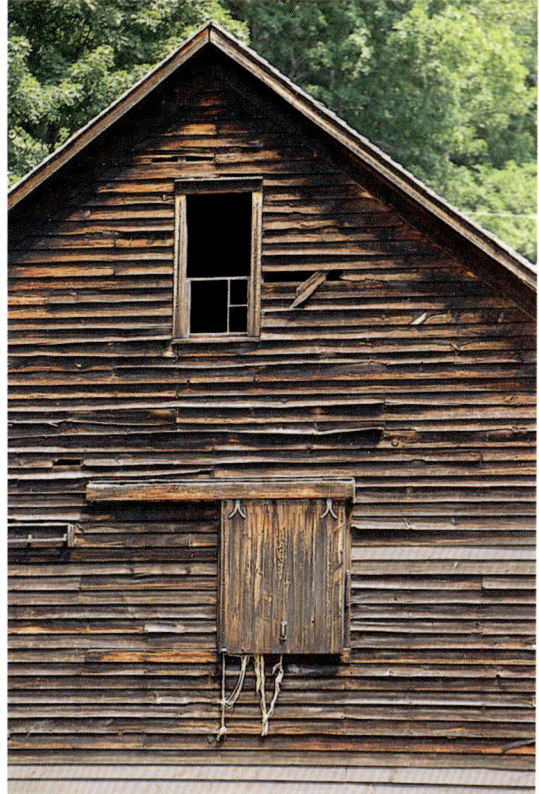

WA-23 装有吊门的风化檐板谷仓
Weathered clapboard barn with hanging door

WA-24 未喷漆的墙板显示了维修的痕迹，双悬窗，美国殖民式样住宅
Unpainted clapboard showing evidence of repair, with double-hung windows, American colonial house

WA-25 装有窗户的喷漆风化木墙板面
Weathered painted clapboard wall with windows

WA-26 传统房屋的彩色封蜡齐平垂直壁板
Stained and sealed flush vertical board siding on traditional house

WA-27 附属建筑物的风化木条镶板
Weathered board-and-batten siding on outbuilding

木制墙面和镶板

WA-28 未喷漆木板板条构成的山墙边缘式墙体、壁板和山墙窗户
Gable-end wall with unpainted board-and-batten siding and gable window

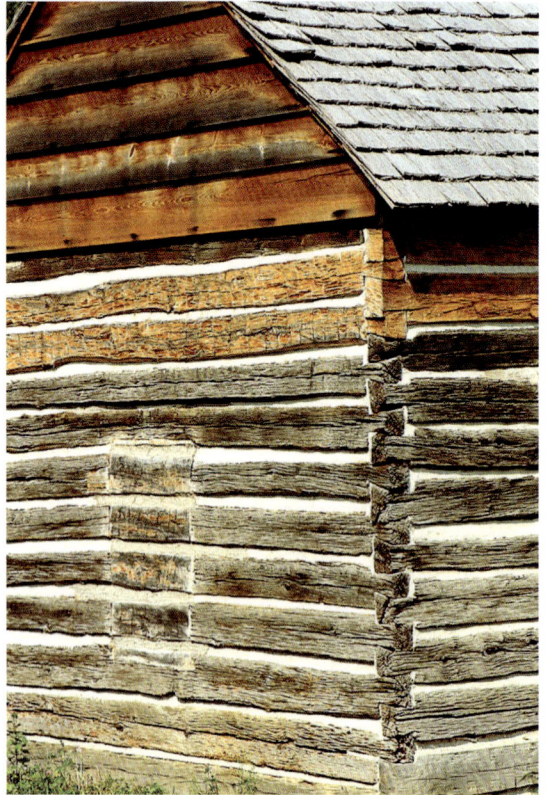

WA-29 楔形榫的拐角和水泥添缝，原木房屋
Dovetail corner and mortar chinking, log house

WA-30 以菱形窗户和垂直扇形镶板为收尾的喷漆垂直风化木板谷仓
Weathered painted flush vertical board cladding on barn gable end with diamond-shaped window and vertical scallop-edged siding

WA-31 装有三角楣饰上盖的窗户和门，美国西部沙龙的外立面
Weathered flush vertical board siding with pediment-capped windows and door, American western saloon facade

木制墙面和镶板

WA-32 被线角分开的喷漆壁板，从顶部到底部：扇形的木质墙面板；板和板条；墙板；还有带有锯齿图案的山墙封檐板
Painted siding divided by molding. Top to bottom: scalloped wood shingles; board-and-batten; clapboard. Bargeboard with saw-tooth motif

WA-33 带有玫瑰花窗和尖顶窗的板和板条镶板；两坡屋顶的门楣入口，卡彭特哥特式教堂
Board-and-batten siding with rose window and lancet windows; jerkinhead-roofed entry, Carpenter Gothic church

WA-34 角塔上的喷漆墙板和大块檐口托饰，意大利式房屋
Painted clapboard and block modillion cornice on corner tower, Italianate house

WA-35 檐板山墙，装有百叶山墙窗、盲窗、扇形倾斜饰边配件和人字形门廊，希腊复兴式房屋
Clapboard end-gable wall with shuttered gable window, blind window, scalloped rake trim, and pedimented porch, Greek-revival house

木制墙面和镶板

WA-36 被梁端的细部节点托起的复斜屋顶上的风化原木墙板
Weathered untreated clapboard on gambrel-end wall rising from corbel detail

WA-37 喷漆的殖民地风格墙板房的宽边角板和一、二层之间的线脚
Painted colonial clapboard house with wide coner board and a jetty between the first and second stories

WA-38 装有角板和波浪形配饰的单坡顶房子上的喷漆舌榫镶板
Painted tongue-and-groove siding on shed-roofed house with corner board and rake trim

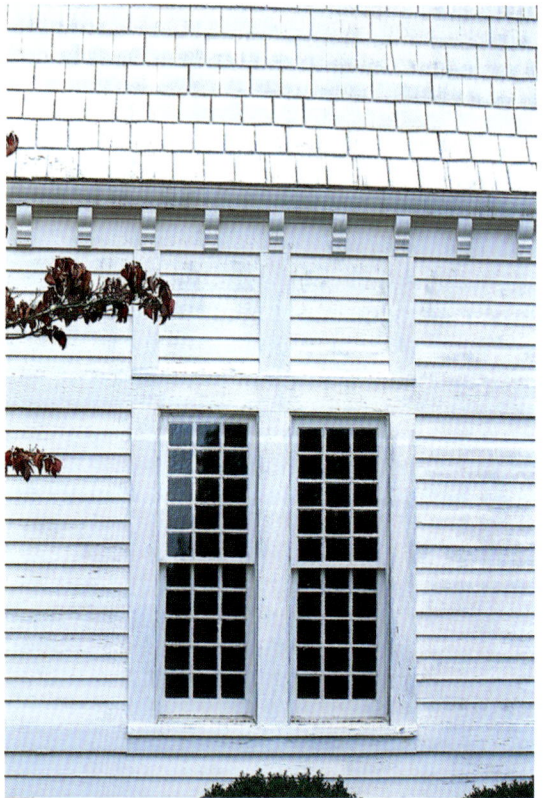

WA-39 由边饰、双悬窗、檐口托饰和喷漆独立屋顶分隔开的喷漆壁板
Painted clapboard siding divided by wide trim, double-hung windows, modillion cornice, and painted shingled roof

木制墙面和镶板

WA-40 以交替斜纹方式形成的舌榫嵌板，印度尼西亚挑空而建的房屋
Tongue-and-groove panels in alternating diagonal patterns, Indonesian house on stilts

WA-41 由弦切杉木板填实的木质框架，带有雨水斗的檐槽
Timber frame with infill of plain-sawn cedar planks; gutter with rainwater head

WA-42 壁板上带装饰框的格式化半木墙
Stylized half-timber wall with decorative framing over clapboard infill

WA-43 装有水平木条的弦切垂直木板搭建在规则的石头地基上；木头底座的支柱
Plain-sawn vertical planks with horizontal batten on shaped stone foundation; timber post with wood base

木制墙面和镶板

WA-44 碎石地基上装有垂直原木的风化粗加工弦切木板墙
Weathered unfinished plain-sawn planks with vertical undressedtimbers on rubble foundation

WA-45 带弦切木板的厚木框架和可以移动的竹条栅栏
Timber-frame with plain-sawn planks and split-bamboo removable fence

WA-46 装有滑动门和纸糊窗户的厚木板和板条填实木框架
Timber frame with infill of planks and battens with sliding doors and paper-covered windows

木制墙面和镶板

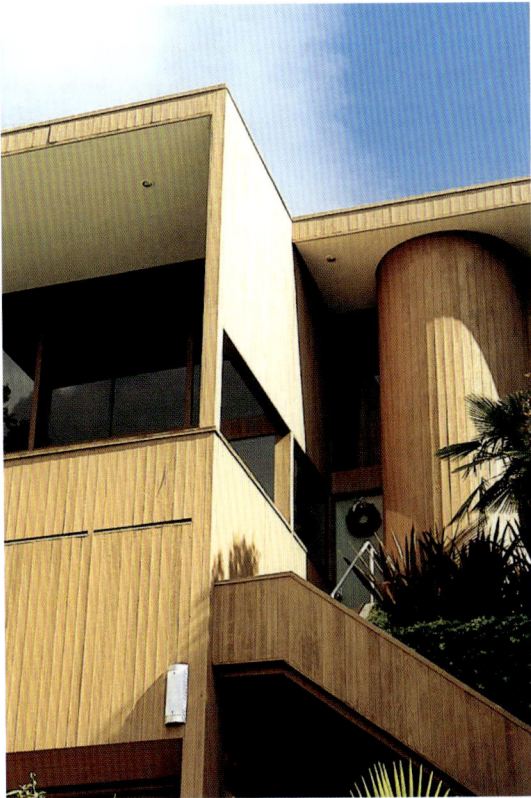

WA-47 平面和曲面上的齐平木条镶板，现代房屋
Flush board siding on flat and curved surfaces, modern house

WA-48 齐平木条镶板，现代商业大厦
Flush board siding, modern commercial building

WA-49 带有重叠连接拐角的防御形式的收分墙
Fortification-style battered wall with lapped-joint corners

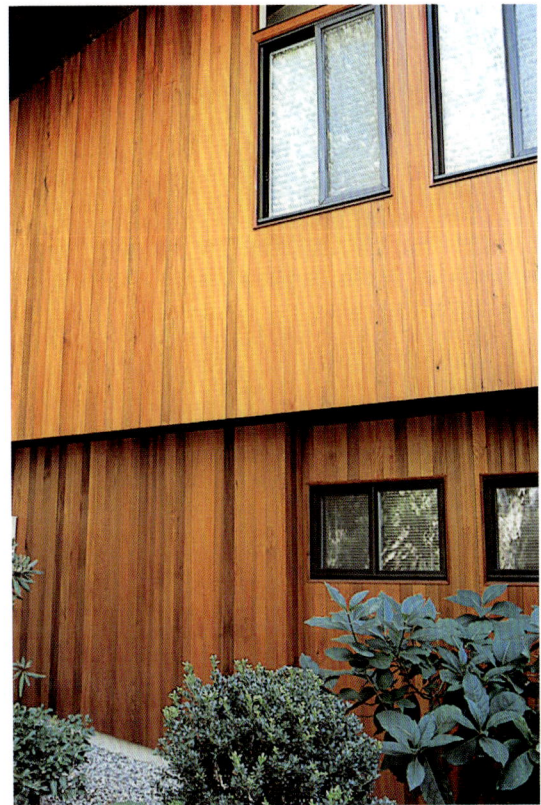

WA-50 彩色封蜡垂直壁板，二层出挑，金属框架窗，现代房屋
Stained and sealed vertical siding with second-floor jetty and sliding metal-framed windows, modern house

木制墙面和镶板

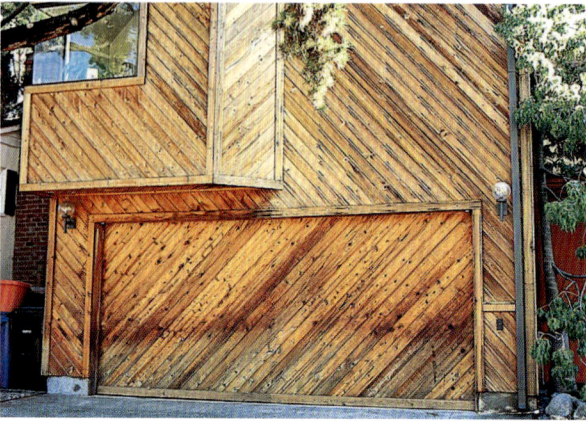

WA-51 带有平边配饰和飘窗的斜纹镶板，现代房屋
Diagonal siding with flat edge trim and corner window, modern house

WA-52 装有拐角凉棚窗户的水平染色接合镶板，现代房屋
Horizontal stained shiplap siding with corner awning windows, modern house

WA-53 带倾斜板的水平和垂直木镶板同斜屋顶成一条直线，现代房屋
Horizontal and vertical board siding with angled siding aligned to roof rake, modern house

WA-54 粗石地基上带有角饰和地基配饰的喷漆垂直板条镶板
Painted vertical board-and-batten siding with corner and base trim on fieldstone foundation

WA-55 单板房的喷漆垂直格子遮蔽门廊
Painted horizontal-lattice screened porch on shingle-clad house

木制墙面和镶板

WA-56 装有通风口的层列碎石中的劈裂面粗糙方形砂岩
Split-faced roughly squared sandstone in coursed rubble with ventilation slit

WA-57 顺砖砌合中切割完整的火山岩或浮石
Sawn-finish volcanic or pumice stone in running bond

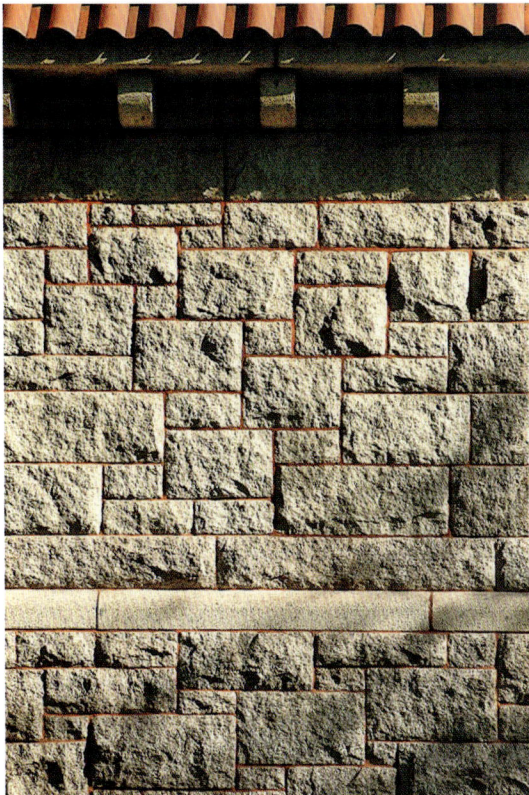

WA-58 带光面水平分隔带的乱砌砖面中的劈裂面花岗岩
Split-faced granite in random ashlar with smooth-faced water-table course, tinted mortar

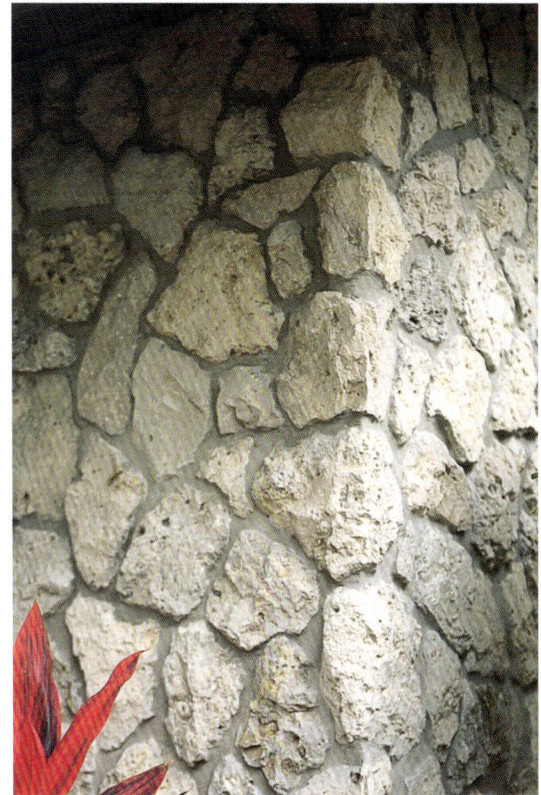

WA-59 镶嵌式化石拐角细部
Corner detail of fossil rock in mosaic bond

石材墙面

WA-60 带有水泥砖石结构窗口的层列碎石墙中的沙滩卵石
Beach pebbles in coursed rubble with masonry screen window

WA-61 人字形图案中的方块劈裂面花岗岩
Squared split-faced granite in herringbone pattern

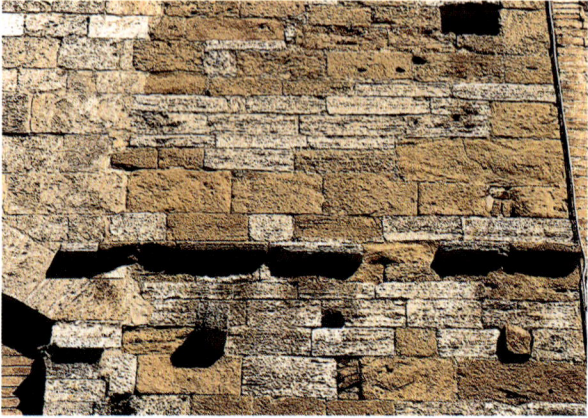

WA-62 层列方砖的古罗马石灰华墙
Ancient Roman travertine wall in coursed ashlar with sandstone repairs

WA-63 带有镶嵌式锤琢面砂岩的水池墙与由沙滩卵石和粗面方石构成的水线层
Pool wall in hammered-finish sandstone with mosaic bond and water-line courses of roughly squared stone and beach pebbles

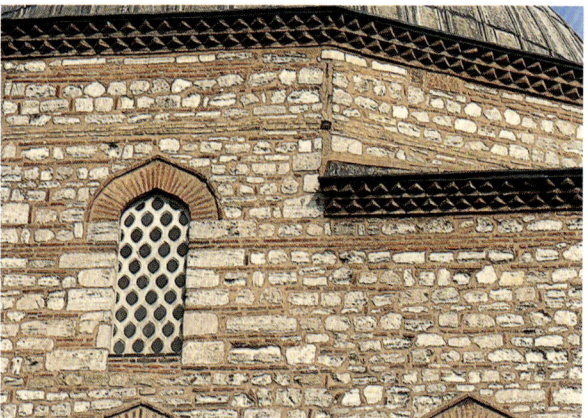

WA-64 带有砖层、浅色灰浆、规则楔形石和葱形拱窗上的水泥砖石的方形碎石清真寺
Squared rubble with brick course, tinted mortar, dimension-stone quoins, and masonry screen in ogee-archwindow, mosque

WA-65 分割的岩面的砂岩；用刮浆随机层砌的琢石
Split- and rock-faced sandstone; shelf course, random ashlar with raked mortar

石材墙面

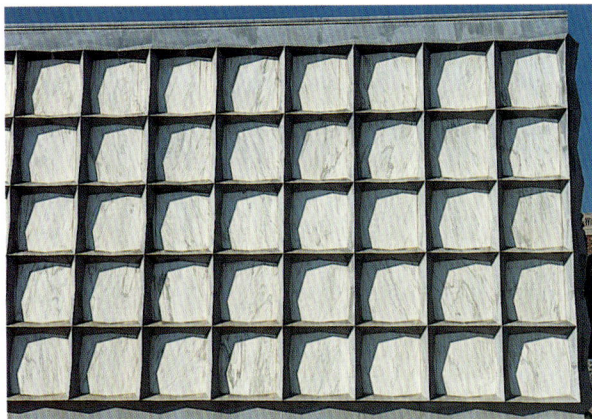

WA-66 用弗蒙特大理石填充的混凝土框架上的花岗岩贴面板，现代商业大厦
Granite veneer on concrete frame with Vermont marble infill, modern commercial building

WA-67 由灰浆接合琢石和锤面褐色砂石，并带有承雨线脚
Hammered-finish brownstone in coursed ashlar with flush mortar joints and water-table course

WA-68 条纹砂岩，顶部：顺砖砌合，饰有岩面带的细磨表面；底部：顺砖砌合，裂缝面；现代商业大厦
Banded sandstone. Top: running bond, honed-finish with rock-face belt course; bottom: running bond, split-face, modern commercial building

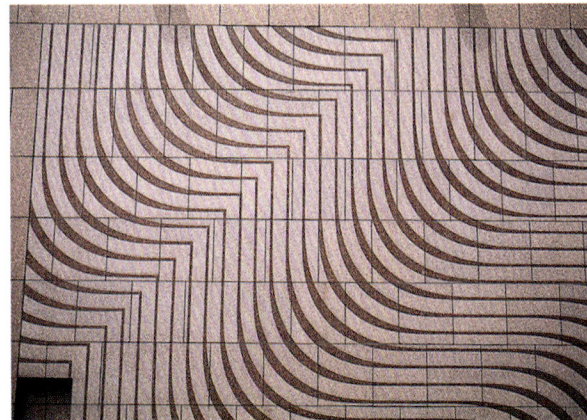

WA-69 带有超大图案的磨光发亮的顺砖砌合的粉色花岗岩
Pink granite in running bond with polished and flamed finishes forming supergraphic, modern commercial building

WA-70 顺砖砌合上的白色和绿色条纹大理石，现代商业大厦
Banded white and green marble in running bond, modern commercial building

WA-71 大厅墙面上顺砖砌合上带有维特鲁威风格的卷形中楣
Siena marble in running bond with Vitruvian scroll frieze on hallway wall

摄影：The Preservation Society of Newport County

石材墙面

WA-72 裸露的碎石结构墙体上层列琢石砂岩墙面
Coursed ashlar sandstone wall in ruins revealing structural rubble wall

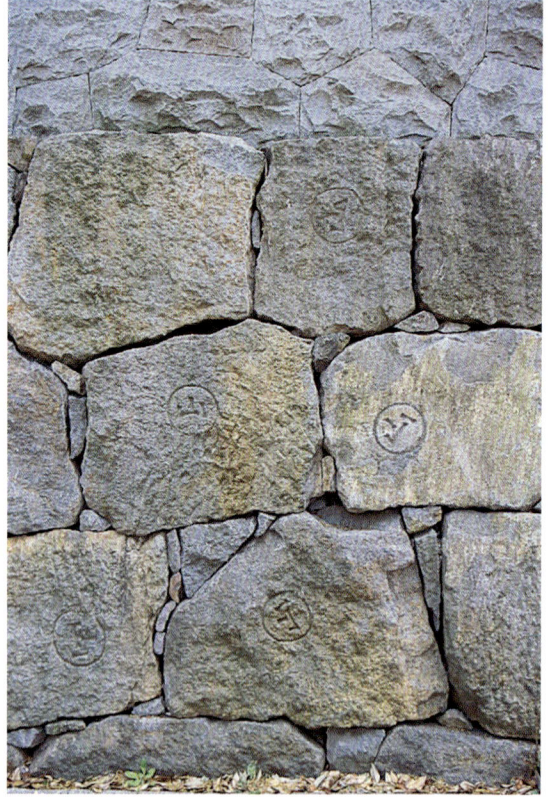

WA-73 带有碎石片和岩面柱冠的层列碎石
Coursed rubble with gallets and rock-face cap

WA-74 顶部：镶嵌砌合的粗石；底部：层列的海卵石
Top: fieldstone rubble in mosaic bond; bottom: coursed rubble of beach pebbles

WA-75 带有砂面的层列琢石石灰墙面，有水侵风化破坏现象
Coursed ashlar limestone with sand finish, water-damage weathering

石材墙面

WA-76 带有碎石片的巨大碎石，花园围墙
Massive rubble with gallets, garden wall

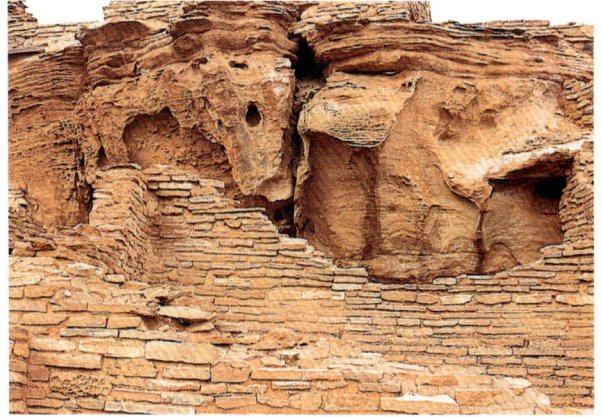

WA-77 同自然石墙相结合的随机堆砌的琢石面，其中有粗大的方形砂岩
Roughly squared sandstone in random ashlar attached to natural stone wall

WA-78 顺砖砌合中分割面和粗面墨西哥火山石，人行桥墙面
Split-faced and rusticated Mexican volcanic stone in running bond, footbridge wall

WA-79 在低墙的上方和高墙顶部带有方形宽石阶的卵石墙
Rubble walls of cobbles with wide squared stone step capping lower wall and coping on upper wall

WA-80 有支柱、壁龛和壁龛下贯石的层列琢石墙面上的石头，巴厘岛
Stone in coursed ashlar with bracket, niche, and perpend under niche, Bali

WA-81 有壁柱和石板遮檐的层列碎石风化墙
Coursed rubble drywall with pilaster and flagstone coping

石材墙面

WA-82 交替岩面的层列和切割花岗岩的垛式砌合中的入口细
部；倾斜拐角接合处
Detail of doorway reveal in stack bond with alternating courses of
rock-face and sawn granite; mitered corner joints

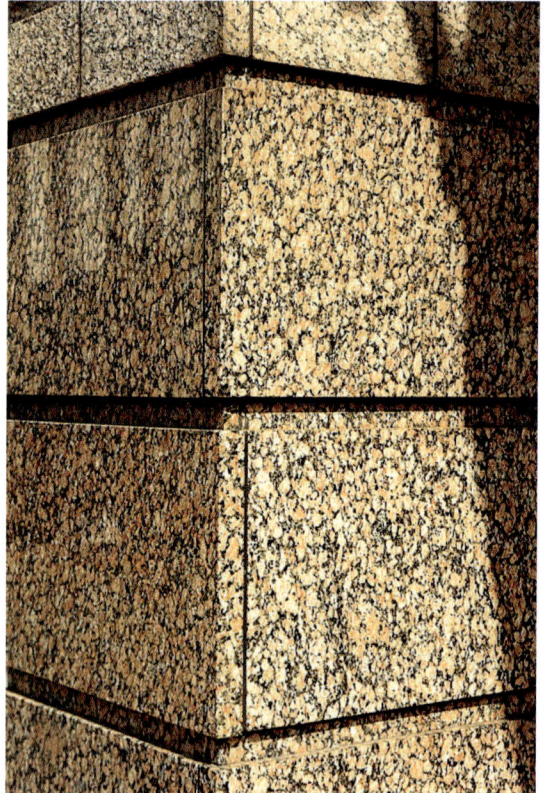

WA-83 带有重叠拐角接缝的光面花岗岩贴面板的拐角细部
Corner detail of polished granite veneer with lapped corner joints

WA-84 光面彩色装饰花岗岩贴面板的边缘细部
Edge detail of polished and honed polychrome granite veneers

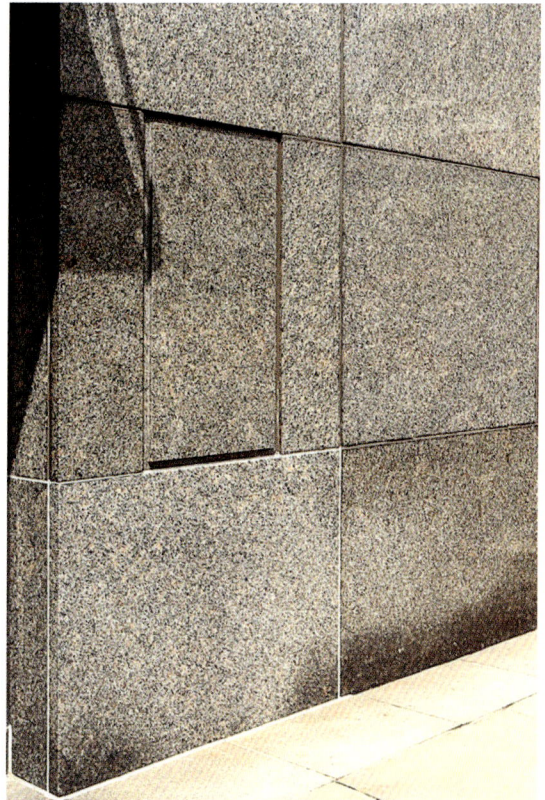

WA-85 薄浆接合的垛式砌合中的细磨面花岗岩
Honed granite veneer in stack bond with grouted joints

石材墙面

WA-86 人字形图案下的砂岩和石英岩层列琢石墙面，并带有印度石灰石装饰柱冠
Sandstone and quartzite in coursed ashlar below herringbone pattern; Indiana limestone ornamented cap

WA-87 石灰岩墙面，左：带有平拱和齿状上楣的斜截垛式砌合；右：宝石状切割的垛式砌合
Limestone walls. Left: bevel-cut stack bond with jack arch and dentil cornice; right: gem-cut stack bond

WA-88 光面花岗岩和大理石，从右到左：以凸形线脚为基础的花岗岩贴面；带有大理石圆凸形线脚装饰的花岗岩底座；光面大理石支柱
Polished granite and marble. Left to right: granite-veneer base with bead molding; granite plinth with marble ovolo molding on base; polished granite engaged column

石材墙面

WA-89 实心花岗岩基座和花岗岩门框边的凸形线脚的光面花
岗岩饰面
Honed granite veneer with solid granite base and bead molding
beside granite door casing

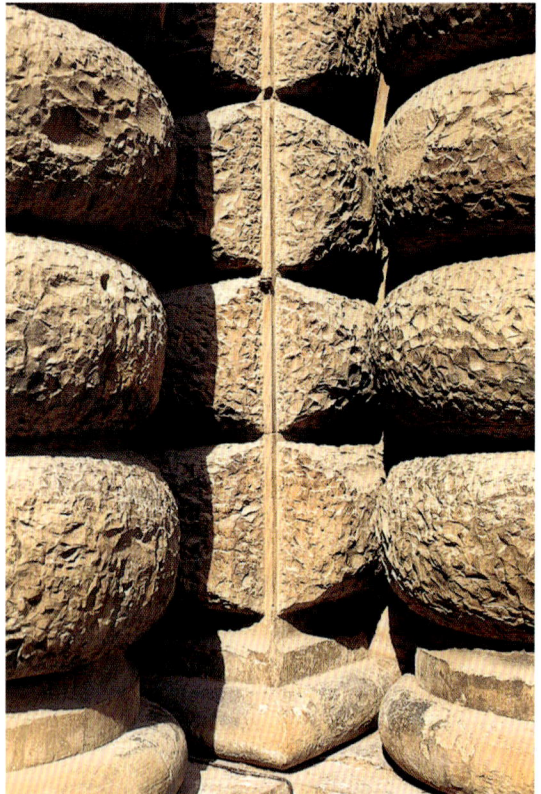

WA-90 两个带状附墙圆柱间的粗石壁柱的拐角细部
Corner detail of rusticated stone pilaster between banded
engaged columns

WA-91 随机堆砌的琢石墙面中的粗面方形石头，在带有嵌石
板拱石的混合拱门中
Roughly squared stone in random ashlar wall with flagstone voussoirs in compound arch

WA-92 锯琢石砂石层列，表面有蠕虫形图案的楔形石
Sawn ashlar sandstone courses, quoins with vermiculated finish

石材墙面

摄影:Lockwood-Mathews Mansion Museum

WA-93 随意堆砌的琢石墙面上的锯岩，扁平花岗岩窗框
Sawn granite in random ashlar with tooled quoins; flat granite window and base trim

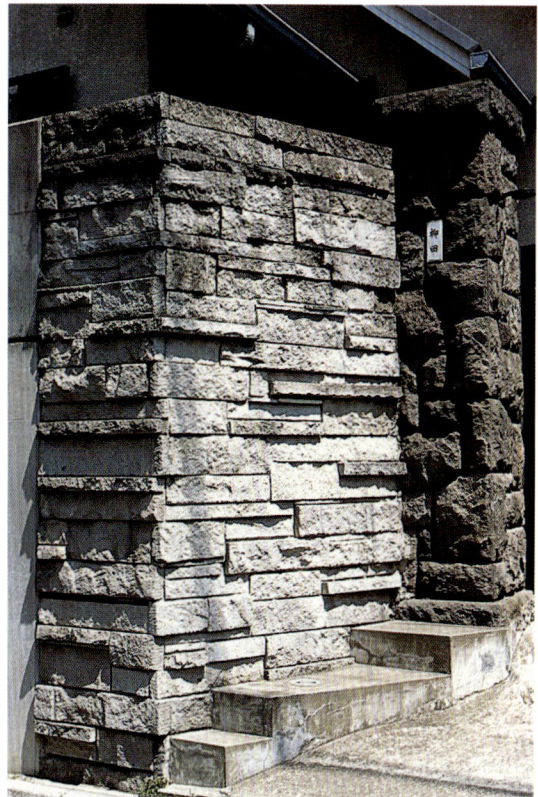

WA-94 岩石上的花岗岩分割墙面，带有正方形粗面碎石墩的随意琢拼饰墙面板（右）
Split-faced granite wall in shelf, random ashlar with roughly squared rubble pillar (right)

WA-95 分割花岗岩面，有承雨线脚的层列琢石扶墙
Split-faced granite, coursed ashlar buttresses with water-table course

WA-96 有砂面枕梁的层列方形毛面碎石
Split-faced coursed roughly squared rubble with sand-finished corbel

石材墙面

WA-97 顺砖砌合中内部拐角的光磨贴面板，底部近期清洁过
Inside corner of honed-veneer marble in running bond, bottom recently cleaned

WA-98 阿默斯特灰色石灰岩的外部拐角，带有多种形状嵌板的砂石表面
Outside corner probably of Amherst gray limestone, sand-finished in shaped panels

WA-99 割面的外部拐角，凸起砂面楔合石和啮合栏杆的随意拼饰琢石墙面
Outside corner of split-faced, random ashlar with raised sand-finished quoins and engaged balustrade

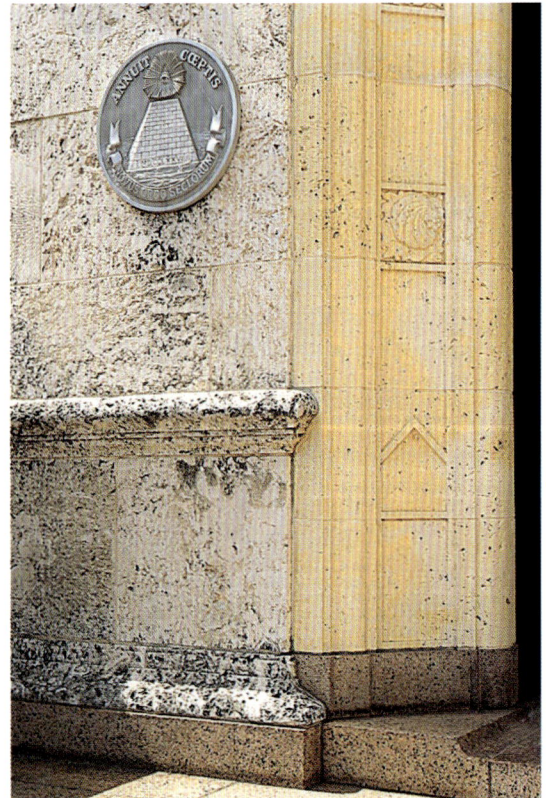

WA-100 有水平腰线和基础线脚的半光面化石门廊
Partially honed fossil-stone doorway with water-table course and base molding

石材墙面

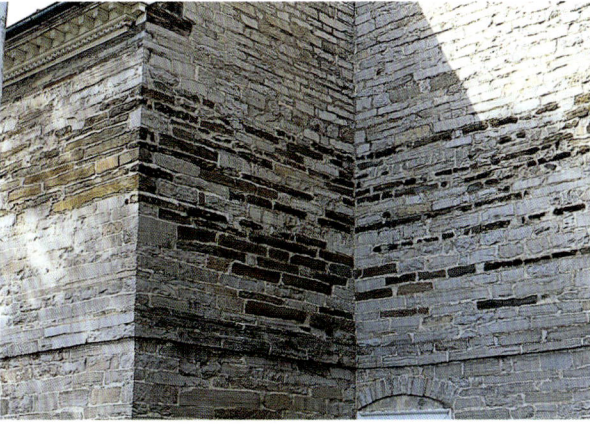

WA-101 方形层列琢石墙面上的内外角
Inside and outside corners in coursed, roughly squared ashlar

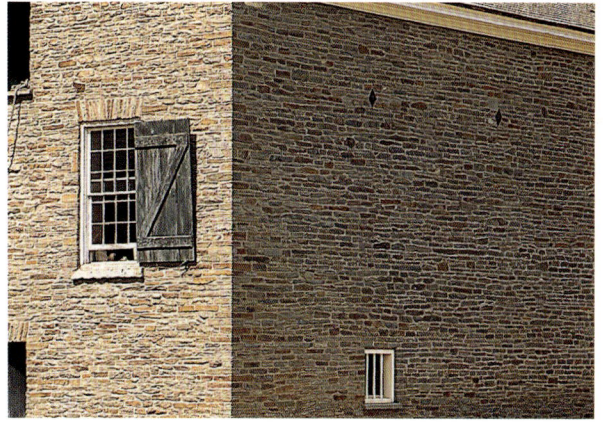

WA-102 嵌有锚型铁饰的随意层列琢石墙面的拐角；装有平拱的窗户
Corner of coursed, random ashlar with anchor irons; window with jack arch

WA-103 有线脚层和框架嵌壁式嵌板的层列琢石面中的石灰岩，有水蚀风化破坏
Limestone in coursed ashlar with molding course and framed, recessed panels, water-damage weathering

WA-104 楼梯底部的锡耶纳大理石贴面板，带有线脚和壁柱
Siena marble veneer on staircase soffit, molding and pilaster

摄影：The Preservation Society of Newport County

WA-105 有齿状檐口和拱形窗的分离花岗岩墙面：佛罗伦萨式（左和中），弓形（右），下面带有平拱门的入口
Split-faced coursed ashlar granite wall with dentil cornice and arched windows: Florentine (left and center), segmental (right), with jack-arch entries below

石材墙面

35

WA-106 层列碎石的内角角楼，外角和窗户四周嵌有大块楔形石
Inside-corner turret in coursed rubble with dimension-stone quoins on outside corners and around window

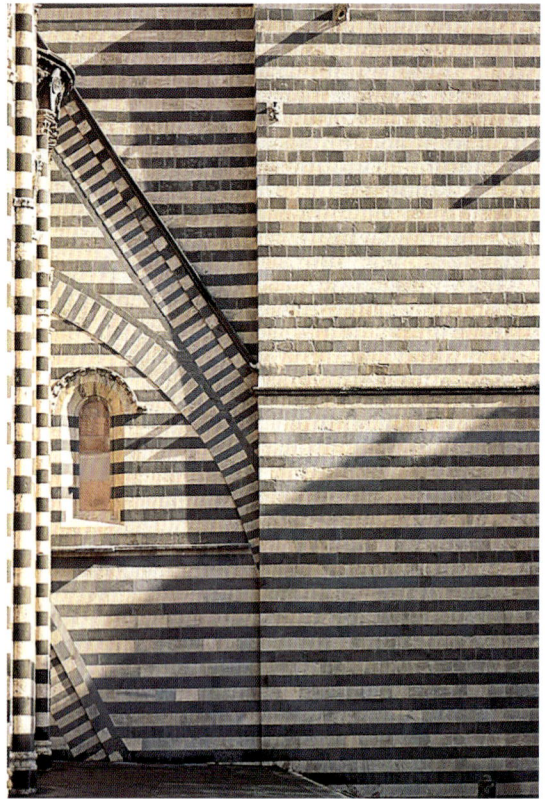

WA-107 配有铲齿拱的交替彩色琢石层中的条纹大理石
Banded marble in alternating polychrome ashlar courses with relieving arch

WA-108 随意的卵石碎块墙面和粗大的棕色楔形石；有隔带以及拱形窗四周的拱石
Random rubble of cobbles with rusticated dimension-brownstone quoins, belt course, and voussoirs around arch window

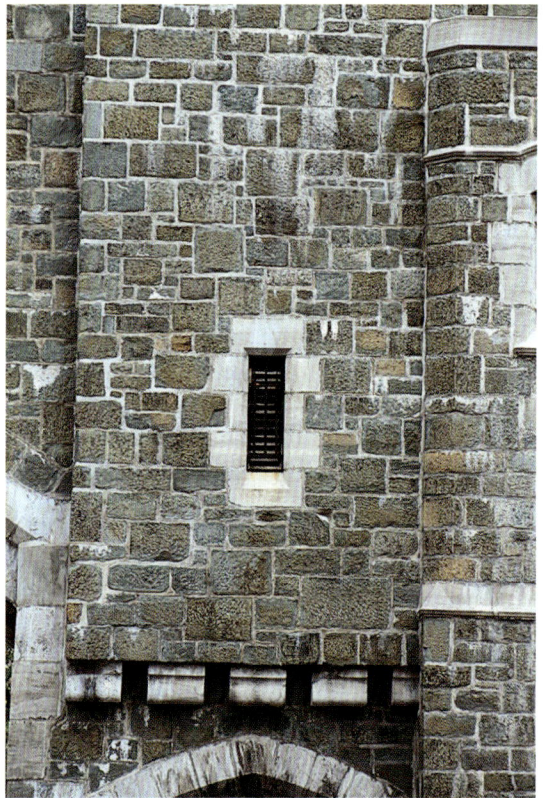

WA-109 花岗岩在带有托臂式出挑和大块窗户楔形石的随意琢石墙面中
Granite in random ashlar with corbelled jetty, dimension-stone window quoins

石材墙面

WA-110 错列搁板层上有石板的入口墙
Entry wall with flagstones in staggered shelf courses

WA-111 层列的方块碎石和人字形图案中的页岩粗石
Shale fieldstone in coursed roughly squared rubble and herringbone pattern

WA-112 有手工面的粗大的顺砖砌合，砂面拱心石；门耳；雕饰带
Rusticated running bond with hand-tooled finish; sand-finish keystone; crossettes; carved ornamented belt course

WA-113 顺砖砌合中的风化砂面石灰岩，梯状拱门中带有涡卷拱心
Weathered sand-finish limestone in running bond, with cartouche keystone in stepped arch

石材墙面

WA-114 部分干净的印第安纳州伯明翰暗花色石灰岩，锯式贴面板，在带有凸窗的随意琢石墙面结构中
Partially cleaned Indiana Birmingham buff limestone, sawn veneer, in random ashlar with oriel

WA-115 有齿状装饰和檐口托饰的层列琢石墙面中的风化石灰岩；支柱和窗框上的雕饰
Weathered limestone in coursed ashlar with dentil and modillion cornice; carved ornament on columns and window frames

WA-116 有凸起花纹的拱石层列琢石墙面中的锯材化石；有凸圆基础线脚的啮合栏杆
Sawn fossil-stone in coursed ashlar with raised voussoirs; engaged balustrade with bead base molding above belt course

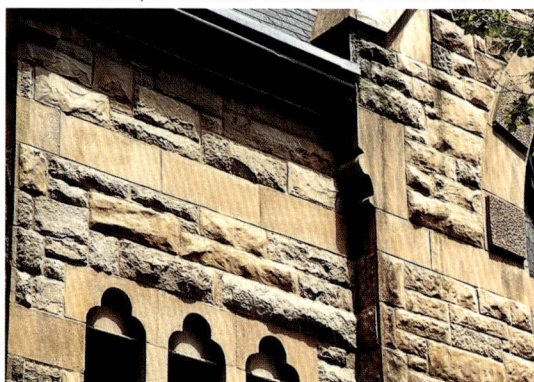

WA-117 自由琢石墙面中的岩面棕石；砂面隔带，三叶草雕饰楣石
Rock-face brownstone in random ashlar; sand-finished belt course, trefoil carved lintel

WA-118 有上楣的顺砖砌合中的石灰岩；雕饰拱腹的上层；各种样式的窗口
Limestone in running bond with cornice; upper story with carved spandrels; varied window openings

WA-119 带有艺术浮雕螺旋形装饰和线脚的砂面石灰岩贴面板
Sand-finish limestone veneer with Art Deco bas-relief cartouche and molding

石材墙面

WA-120 交替大理石彩边贴面板，在带有摩尔封闭拱门和嵌壁式窗户的层叠砌合和顺砖砌合
Alternating polychrome bands of marble veneer in stack and running bonds with Moorish blind arches with recessed windows

WA-121 带有哥特式封闭拱门和啮合支柱的层列琢石墙面结构中的梯墙
Staircase wall in coursed ashlar with Gothic blind arches and engaged columns

WA-122 镶嵌式碎石墙面结构中带有壁龛和石顶的梯墙
Staircase wall in mosaic-bond rubble with nuches and stone coping

石材墙面

WA-123 自由琢石结构中的三角墙；砖石檐口托饰；拱窗四周的楔形石；托臂窗台（左）

Gable-end wall in random ashlar; block modillion cornice; quoins around arched windows; corbelled window sill (left)

WA-124 以层列毛面方砖琢石墙面结构为基础的倾斜墙

Battered-wall base in coursed roughly squared ashlar

WA-125 有粗面顺砖砌合的花岗岩；层叠砌合侧门入口的新科林斯式的壁柱

Granite in running bond with rustic accents; neo-Corinthian pilasters in stack bond flanking entry

石材墙面

WA-126 啮合化石墙和层列琢石墙面结构中的楼梯
Sawn-fossil-stone wall and staircase in coursed ashlar

WA-127 饰有宝石状砖块的彩色条纹花岗岩贴面墙的拐角细部
Corner detail of polychrome banded granite veneer with gem-cut decorative block

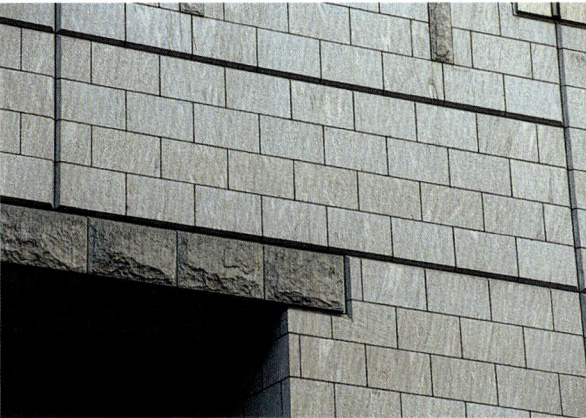

WA-128 岩面花岗岩门廊楣石，其琢石墙面结构中有对比鲜明的V字形接合的啮合花岗岩嵌板
Rock-face granite doorway lintel with contrasting sawn-granite panels in running bond with wide V-shaped joints

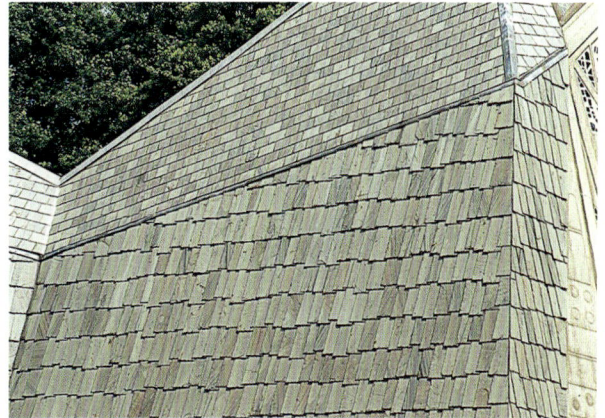

WA-129 页岩顶和骨架外墙的内外角，现代商业大厦
Inside and outside corners of slate roof and cladding, modern commercial building

WA-130 墙面和支柱上的端头拼接的石灰华贴面板的内外角，现代商业大厦
Inside and outside corners of end-matched travertine veneer on walls and post, modern commercial building

石材墙面

WA-131 倾斜结合的花岗岩贴面板外角；窗和水平隔带，现代商业大厦
Outside corner of granite veneer with raked joints; belt courses at window and jetty, modern commercial building

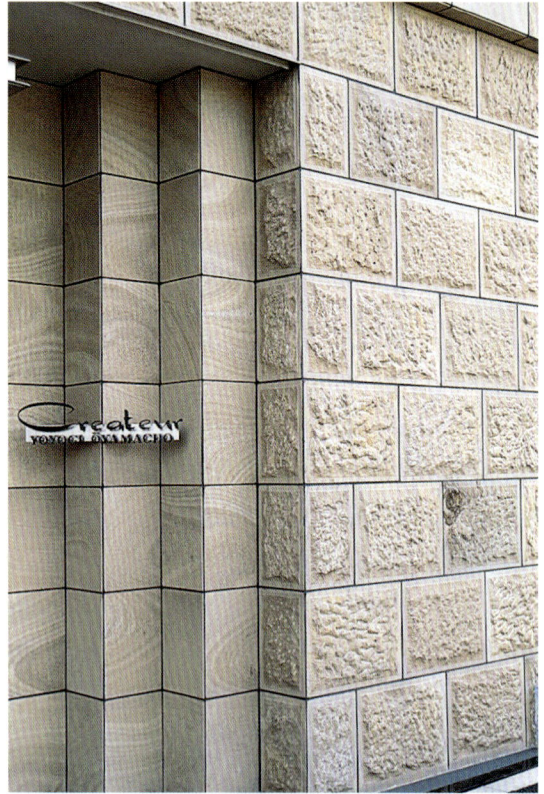

WA-132 带展示的粗面顺砖砌合的拐角；入口处层叠砌合中的砂面壁柱片，现代商业大厦
Corner of rusticated running bond with reveal; sand-finish pilaster strips in stack bond on entry, modern commercial building

WA-133 搁板上的分隔面花岗岩拐角，有金属边的顺砖砌合，现代商业大厦
Corner of split-face granite in shelf, running bond with metal banding, modern commercial building

WA-134 带有灌浆接合处和窗口的层叠砌合锯式花岗岩贴面板的内角，现代商业大厦
Inside corner of honed-granite veneer in stack bond with grouted joints and window reveal, modern commercial building

石材墙面

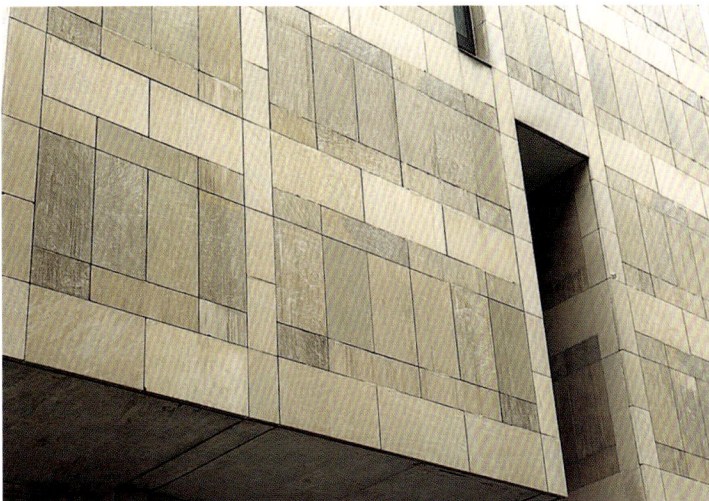

WA-135 被对比鲜明的层叠和顺砖砌合层框起来的石灰石贴面嵌板，现代商业大厦
Limestone-veneer panels framed by contrasting stack-and-running-bond courses, modern commercial building

WA-136 连接在一起的光面花岗岩贴面板，拱肩与亚光面花岗岩和带形窗交替出现，现代商业大厦
Banded polished-granite-veneer spandrels alternating with honed granite and ribbon windows, modern commercial building

WA-137 窗户四周带有凸出图案的层叠砌合亚光面花岗岩贴面板，现代商业大厦
Honed-granite veneer in stack bond with raised pattern around windows, modern commercial building

石材墙面

WA-138 立体构架的花岗岩窗户和花岗岩贴面板形成鲜明对比，现代商业大厦
Granite veneer with contrasting granite window inserts surmounted by space frame, modern commercial building

WA-139 带有突出窗台和岩面装饰块的顺砖砌合的石块，现代商业大厦
Stone in running bond with projecting window sills and rock-faced decorative blocks, modern commercial building

WA-140 带有色彩的砂浆砌合的花岗岩，固定窗户，现代商业大厦
Granite with tinted mortar, fixed windows, modern commercial building

WA-141 彩色裙墙上安有齐平裱贴固定窗的光面花岗岩贴面墙，现代商业大厦
Polished granite veneer with flush-mounted fixed windows in polychrome curtain wall, modern commercial building

石材墙面

WA-142 配有层列砌合的砂岩层列琢石墙面结构框，齐平的大理石贴面板，现代商业大厦
Sandstone coursed-ashlar frame with stack-bond inset, flush marble-veneer base, modern commercial building

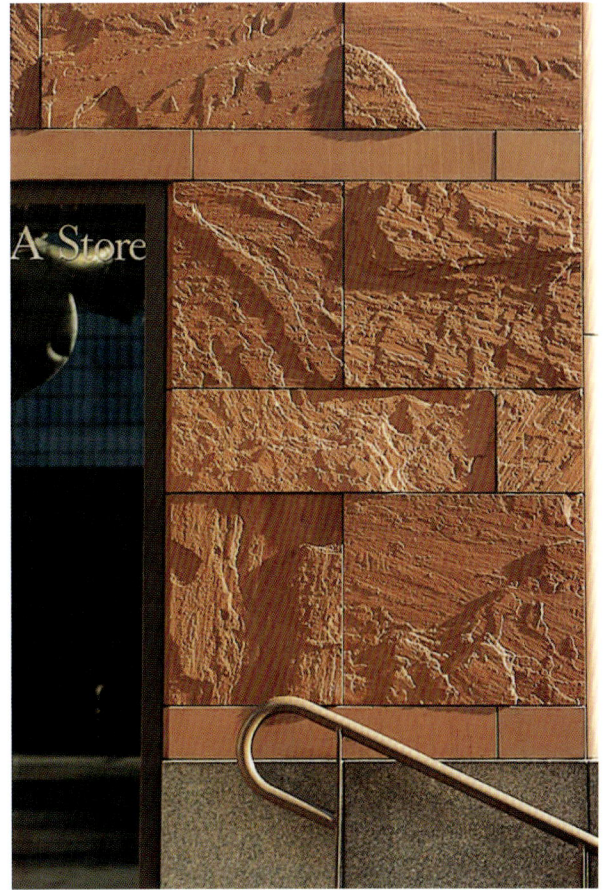

WA-143 自然裂口面的红砂岩，门和黑色花岗岩基座上的有砂面隔带的层列琢石墙面，现代商业大厦
Red sandstone with natural cleft face, coursed ashlar with sand-finish belt courses above door and black granite base, modern commercial building

WA-144 花岗岩，顺砖砌合中装有亚光面和宝石状砖块，现代商业大厦
Granite, with honed and gem-cut blocks in running bond, modern commercial building

石材墙面

WA-145 有壁柱片的彩色镶边的光面贴面板，现代商业大厦
Polished veneer in polychrome bands with pilaster strips, modern commercial building

WA-146 彩色镶边的层叠砌合中的亚光面花岗岩，现代商业大厦
Honed granite in stack bond with polychrome bands, modern apartment house

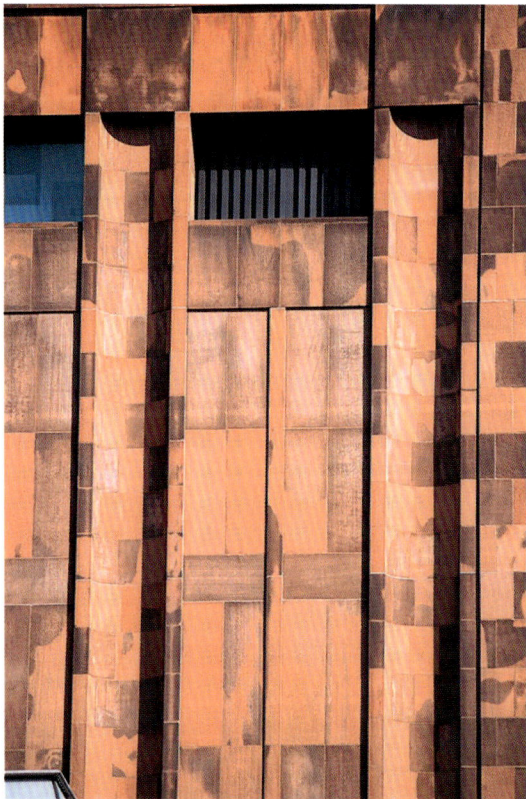

WA-147 对比鲜明的砂面褐色砂石嵌板，现代商业大厦
Contrasting sand-finished brownstone panels, modern commercial building

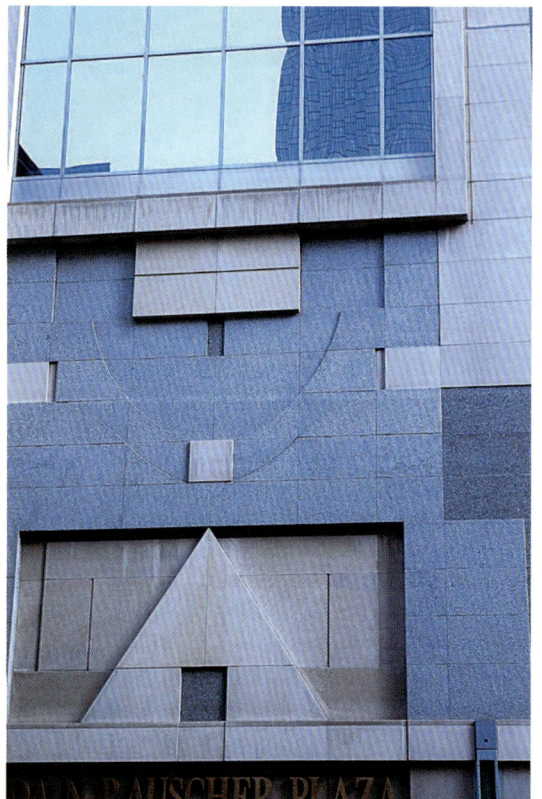

WA-148 抛光面和亚光面花岗岩的彩色石质贴面板拼贴画，现代商业大厦
Polychrome stone-veneer collage of polished and honed granites, modern commercial building

石材墙面

WA-149 左：荷兰式砌合插入嵌板和带有凸出水平嵌板的竖砌砖框架；右：附壁柱顺砖砌合上的白色釉面砖
Left: Flemish-bond inset panel and frame with raised horizontal panel in basket-weave and rowlock bonds; right: white glazed brick on banded pilaster in running bond

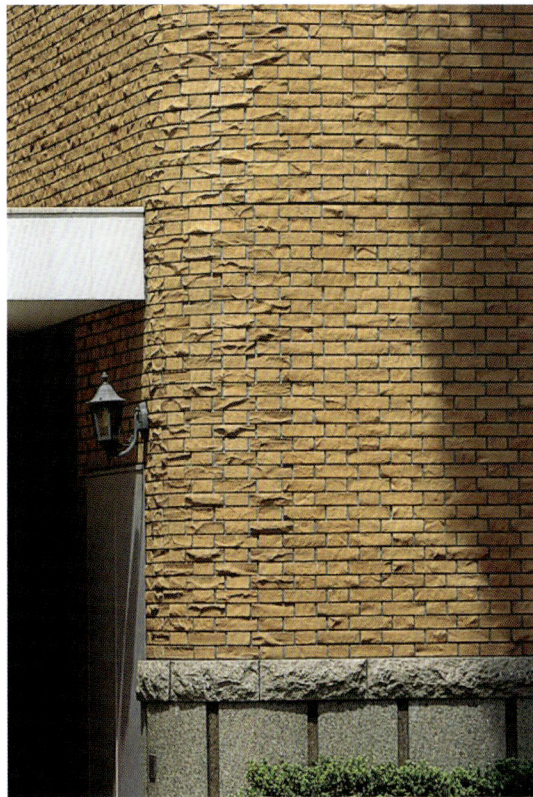

WA-150 荷兰式砌合中的粗面砖，架角，花岗岩分隔带和磨光面底座上带有弯曲拐角
Rusticated brick in Flemish bond, shelf angle, with curved corner above granite rock-face belt course and honed base

WA-151 带有一个倾斜带和立砖压顶的无规则荷兰式砌合
Erratic Flemish bond with a sloping course and rowlock-course coping

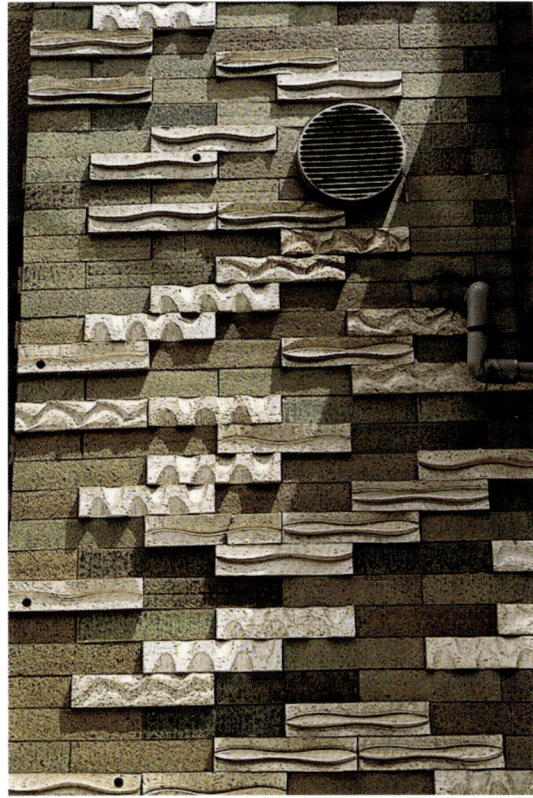

WA-152 定制的雕刻和上釉砖，现代住宅庭院
Custom-sculpted and glazed brick, modern residential courtyard

WA-153 普通外立面砌合，第七层顶梁，修复后的花园围墙
Common bond, seventh-course header, in garden wall with repairs

WA-154 英式砌合中的凸出形状装饰嵌板，18世纪末建筑物
Raised shaped decorative panel in English bond, late 18th-c. building

WA-155 罗马式凸出立砌层和带有石头突额的椭圆形拱，古罗马式墙壁
Roman brick with projecting soldier course and elliptical arch with stone corbels, ancient Roman wall

WA-156 顺砖砌合四周的格子边，混凝土顶石，花园围墙
Basket-weave borders around running-bond; concrete capstone, garden wall

WA-157 立砖压顶的荷兰式砌合的带有凹口的花园围墙
Notched garden wall in Flemish bond with rowlock-course coping

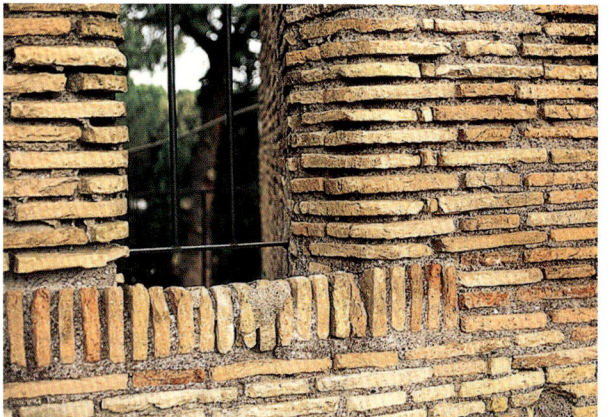

WA-158 以立砌层和楔形砖形成的垛口，古代矮护墙
Roman bricks with soldier course and shaped bricks forming crenel, ancient parapet wall

砖墙面

WA-159 带有楔形石结合隔带的圆线脚的古罗马式砖墙
Ancient Roman brick running-bond wall with stone quoins, banding course, and bead molding

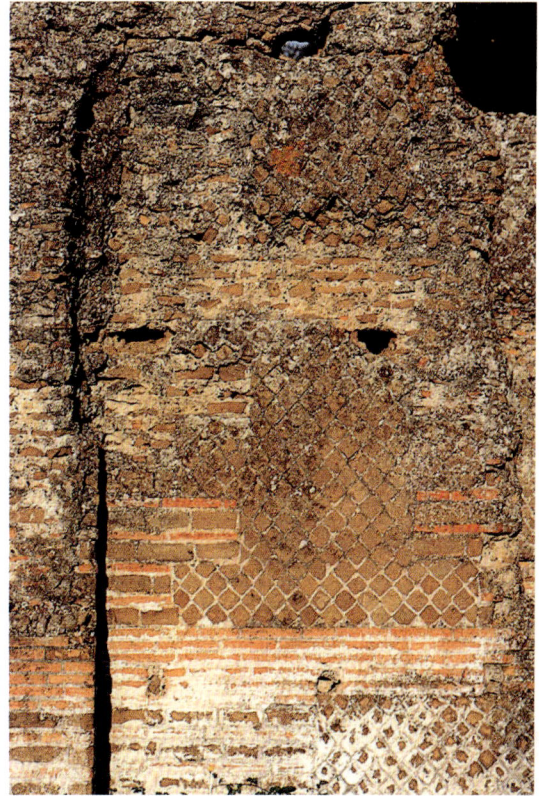

WA-160 有裸露碎石的古罗马式墙，顺砖砌合和对角砌合
Ancient Roman wall with exposed rubble, running and diagonal bonds

WA-161 有封闭半圆拱和立砖砌合枕梁的顺砖砌合的女儿墙
Parapet wall in running bond with blind semicircular arches and corbels below soldier course

WA-162 带有壁柱的罗马砖，古代墙壁
Roman brick with pilaster, ancient wall

WA-163 砖面层由于风化形成结构犬齿隔带的古罗马砖墙
Ancient Roman brick wall with structural dog-tooth course exposed by weathering of brick facing layer

WA-164 在载重拱下，带有回纹装饰的外拱线和嵌壁式拱腹的楔形砖的中世纪拱廊
Medieval arcade under relieving arch in gauged brick with chevron-decorated extrados and recessed spandrels

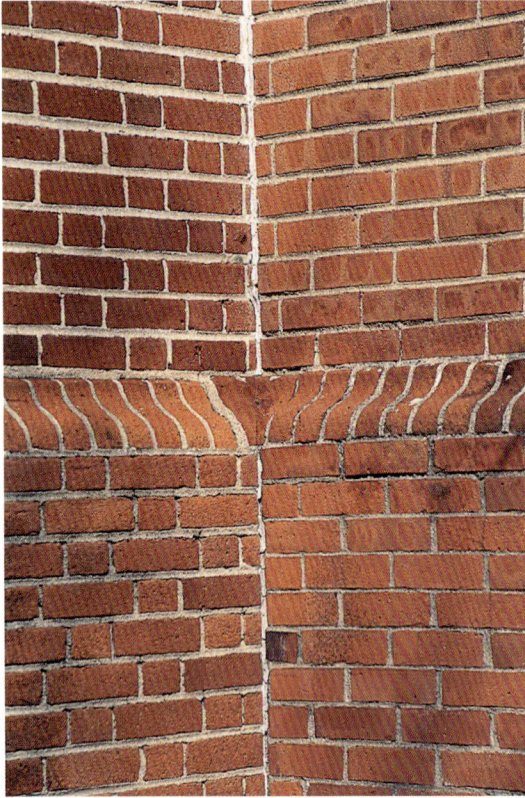

WA-165 有S形曲线水平砖带的荷兰式砌合墙的内角
Inside corner of Flemish-bond wall with ogee water-table course

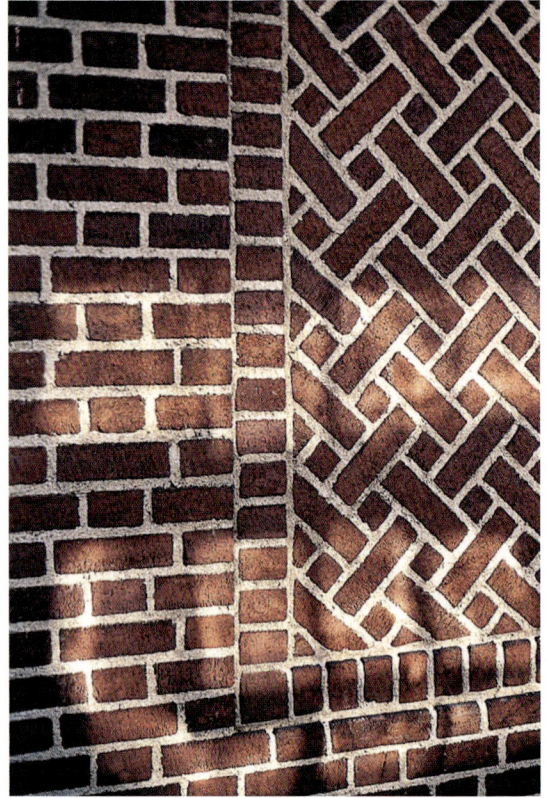

WA-166 多种砌合，从左到右：荷兰式砌合、竖砌合、丁头层砌合框架和席纹图案砌合
Multiple bonds. Left to right: Flemish bond; rowlock and header frame; basket-weave

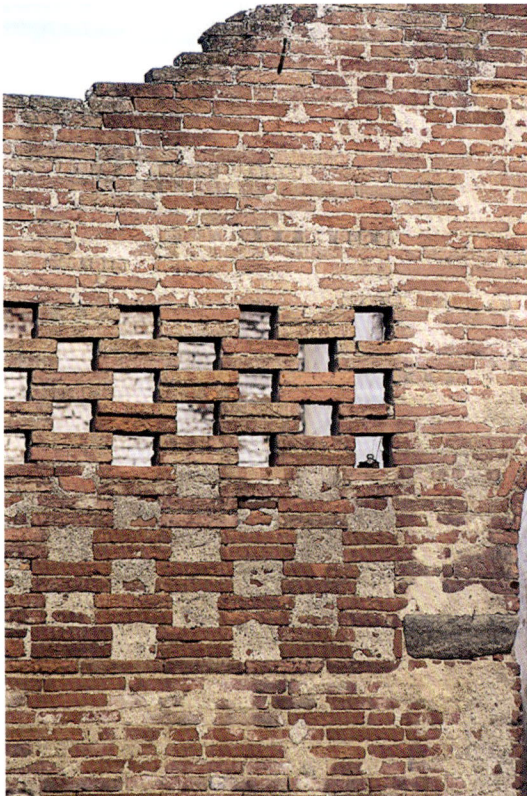

WA-167 罗马砖屏，托斯卡纳农场墙面
Roman brick screen, Tuscan farm wall

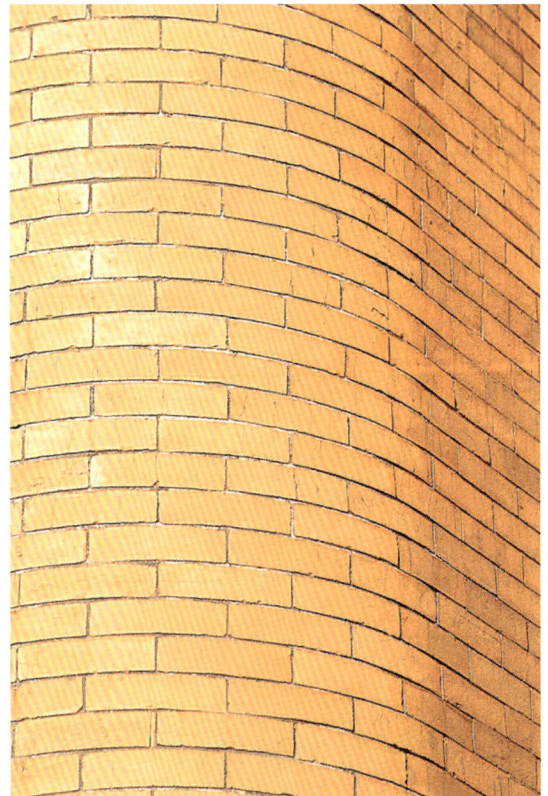

WA-168 弯曲墙面上的顺砖砌合中的弧形砖
Radial brick in running bond on curved wall

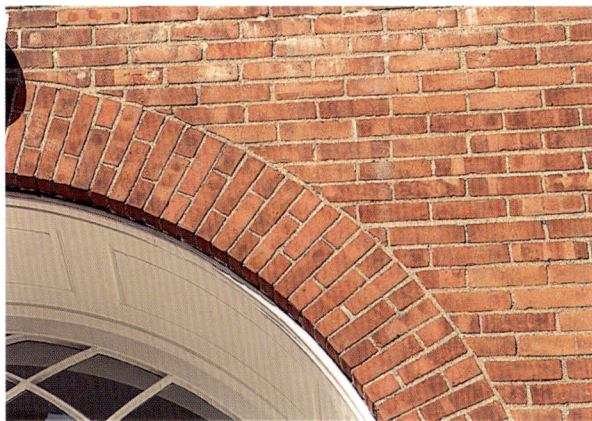

WA-169 环绕顺砖砌合的楔形砖拱细部
Detail of gauged-brick arch surrounded by running bond

WA-170 传统尺寸砖的多种砌合，顶部和右侧：凸出顶梁层；中部：多层框架、立砌和顶梁
Multiple bonds in custom-sized brick. Top and right: raised header courses; center: patterns in stacked stretchers, soldier, and headers

WA-171 特别定制的设计：多种砌合中的多样几何图形
Custom design: multiple geometric patterns in multiple bonds

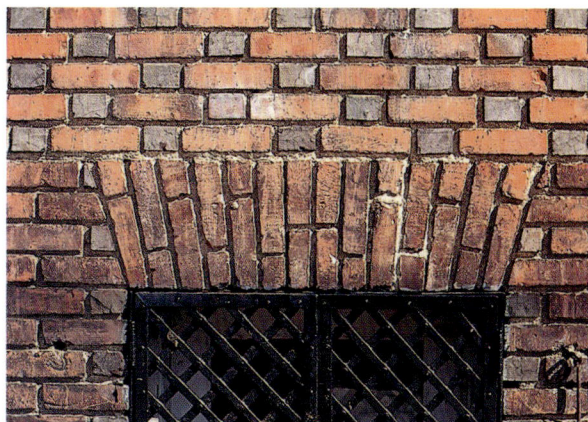

WA-172 交替色的荷兰砌合环绕的平拱
Jack arch surrounded by Flemish bond in alternating colors

WA-173 带有嵌壁式嵌板的英式砌合墙中的平拱窗和对比鲜明的犬齿砖带
Jack-arch window and dog-tooth course of contrasting brick in English bond wall with recessed panel

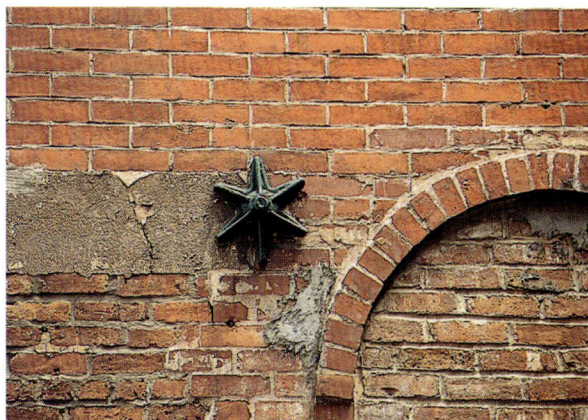

WA-174 带有载重拱的顺砖砌合中的星形的铁锚
Star anchor iron in running-bond brick wall with relieving arch

砖墙面

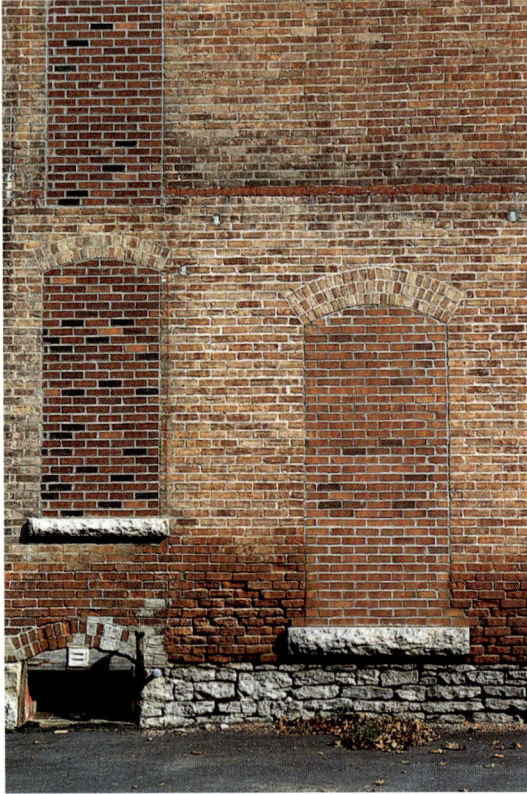

WA-175 粗糙开裂地基与带石窗台的砖填拱形窗上顺砖砌合的风化砖，小巷墙面
Weathered brick in running bond on rough split-faced foundation and bricked-in arch windows with stone sills, alley wall

WA-176 带有檐下齿状线脚造型的喷漆岩面砖、壁柱和带有螺旋装饰的嵌壁式嵌板中的拱形下的双挂窗
Painted rock-face brick with dentil molding, pilasters, and double window under arch in recessed panel with cartouche

WA-177 带彩浆和立砌基础层的顺砖砌合中的罗马砖的现代变化；门和窗户上的百叶
Modern variation of Roman brick in running bond with tinted mortar and soldier base course; louvered shutters on door and window

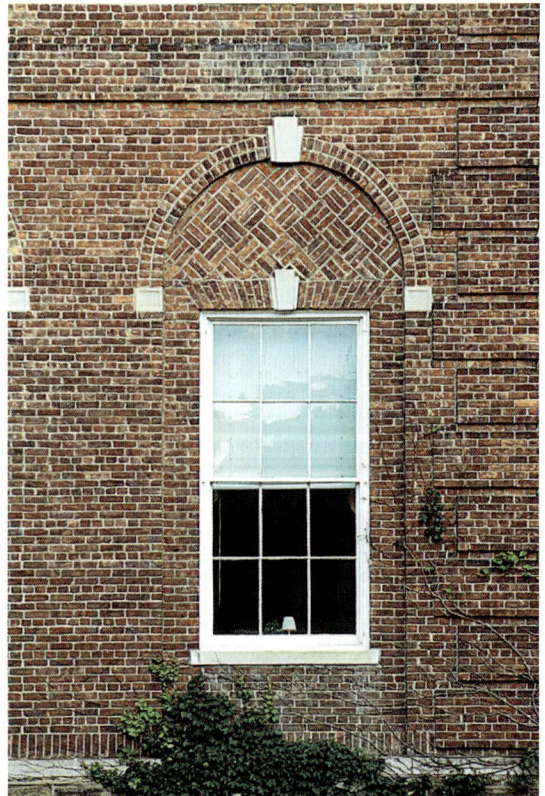

WA-178 第四层顶梁窗上有对角线格子封闭拱和突出楔形石的普通外立面砌合
Common bond with fourth-course header, blind arch of diagonal basket-weave over window, and raised quoin corners

砖墙面

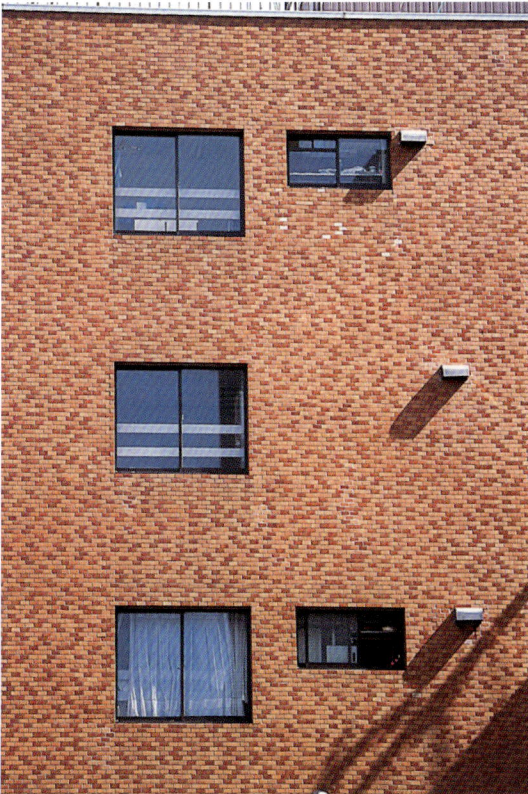

WA-179 彩色砖的斑点效果，现代住宅楼
Mottled effect in polychrome toned brick, modern residential building

WA-180 顺砖砌合中被结合和立砌砖层连接的砖嵌板；窗和底窗，现代商业大厦
Brick panels in running bond articulated by joints and soldier courses; soldier courses top and bottom of windows, modern commercial building

WA-181 带有砖台阶和窗台的弯曲入口处的荷兰式砌合；拐角处有犬齿式砌合变化（最左）
Flemish bond on curved entry with brick steps and window sill; variation of dog's tooth on corner (far left)

WA-182 有六层顶梁的普通外立面砌合；格子图案的嵌壁式嵌板，窗户上檐的立式砖层
Common bond with sixth-course headers; recessed spandrels with checker pattern; soldier course above windows

WA-183 简单砖角，立砖收边的窗口和模仿百叶窗的突出的弯曲线砌合，现代公寓
Rusticated-brick corner; window with rowlock sill, and raised stacked bond imitating shutters, modern apartment house

WA-184 壁柱和嵌板上曲线砌合和水平以及垂直顺砖砌合的特性收集；立砖收边的窗口
Idiosyncratic collection of stacked bond and horizontal and vertical running bonds on pilasters and spandrels; rowlock window sills

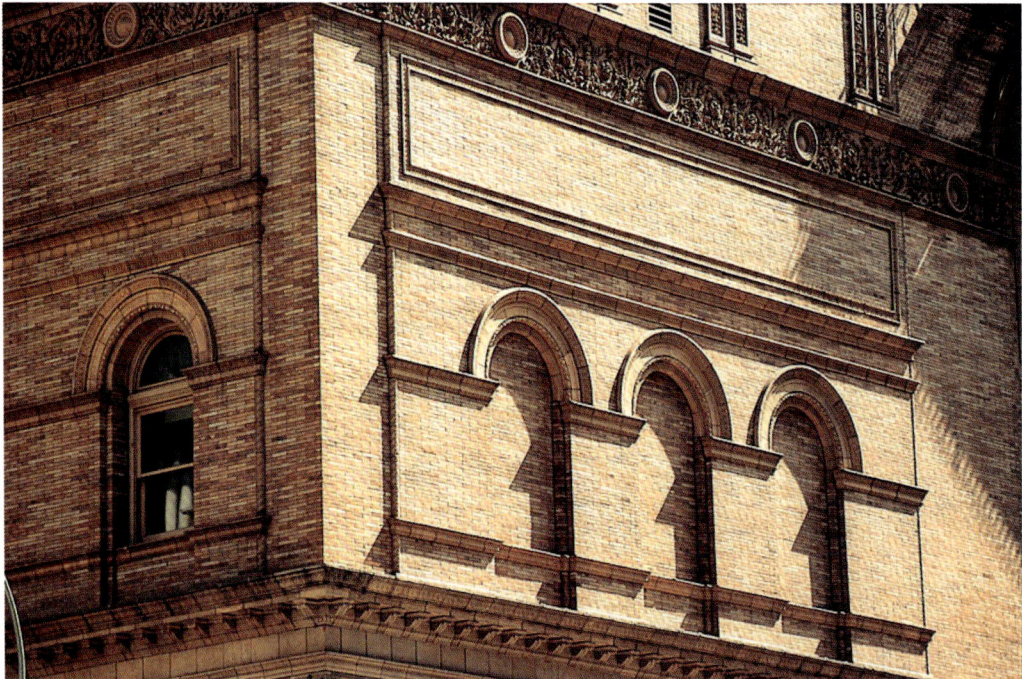

WA-185 上升嵌板上带有赤陶装饰的砖墙封闭拱廊的半圆拱，美术风格商业大厦
Semicircular arches in blind arcade on brick wall with terracotta frieze above raised panel, Beaux-Arts commercial building

墙 面

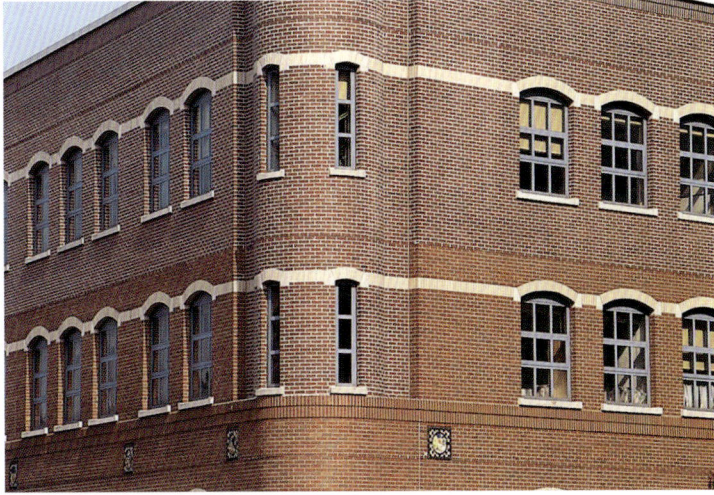

WA-186 对比鲜明的色彩和砖石结构线结合的扇形砖拐角；弧形拱窗，现代商业大厦
Radial brick corner with banding in contrasting colors and masonry; segmental-arch windows, modern commercial building

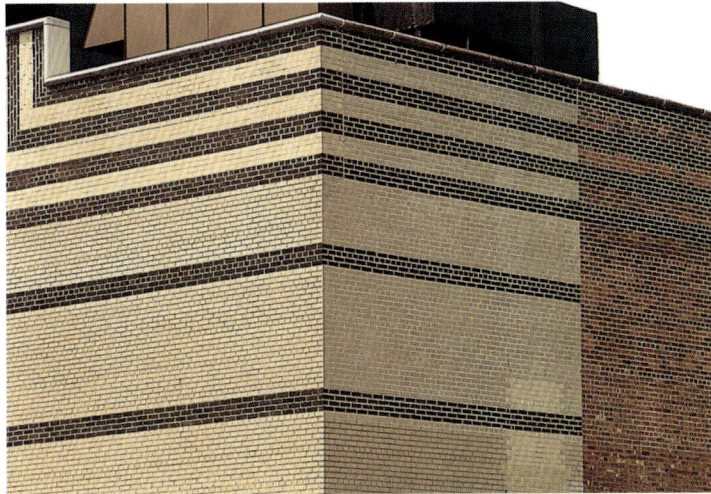

WA-187 拐角细部处的对比色条纹；顺砖砌合，现代商业大厦
Corner detail of banding in contrasting colors; running bond, modern commercial building

WA-188 对比鲜明的砖带和壁柱；标准砖半圆拱，19世纪的公寓
Contrasting brick banding and pilasters; gauged-brick semicircular arches on upper windows, 19th-c. apartment house

砖墙面

WA-189 拐角处如墙屋顶下带有华丽梁托的罗马风格装饰砖结构
Romanesque-style decorative brickwork with ornate corbel on corner below parapet roof

WA-190 弯曲拐角同支架砖层和雕刻砖或建筑陶瓷相结合
Curved corner banded with courses of stretcher bricks and sculpted brick or architectural terracotta

WA-191 顺砖砌合中的镶边细部，立砖和对比色中的立砌砖层，现代商业大厦
Corner detail of bands of running bond, rowlock and soldier courses in contrasting colors, modern commercial building

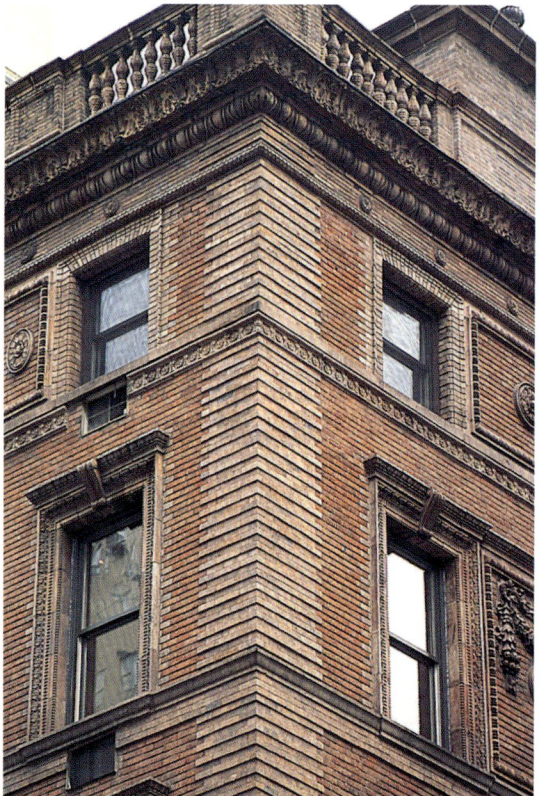

WA-192 砖和装饰模制砖或赤陶的拐角细部，19世纪公寓楼
Corner detail of brick and molded decorative brick or terracotta, 19th-c. apartment house

WA-193 对比色和垂直扩展/密封结合中的砖带拐角细部
Corner detail of brick with banding in contrasting colors and vertical expansion/sealant joints, modern commercial building

WA-194 有犬齿带和突出架的菱形花格图案的定制砌合的拐角细部，现代公寓楼
Corner detail of custom bond with dog-tooth banding and diaper pattern of projecting stretchers, modern apartment house

WA-195 普通外立面砌合的拐角细部，第六层顶梁有犬齿拐角
Corner detail of common bond, sixth-course header, with dog's-tooth corner

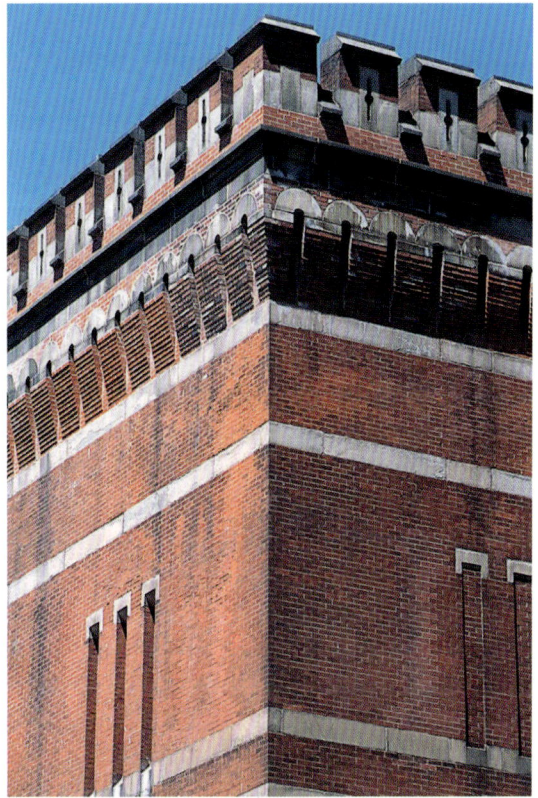

WA-196 有石带的罗马风格的砖砌建筑的拐角细部，矮护墙顶有钝齿
Corner detail of Romanesque-style brick building with stone banding, parapet roof with crenellation

砖墙面

WA-197 带有嵌壁式上下层窗间墙面的玻璃窗，整齐的褐色砂石和楔形石
Brick window-wall with recessed spandrels, brownstone window trim and quoins

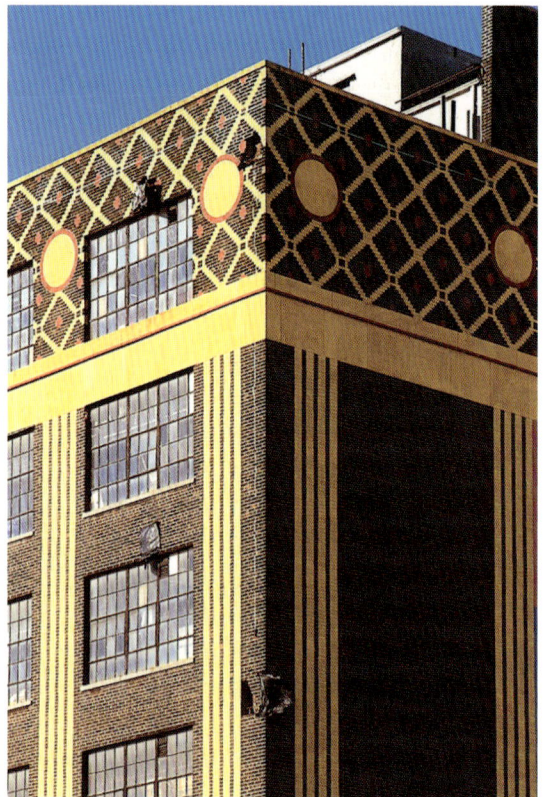

WA-198 装饰砖结构的拐角细部：饰有图形和垂直对比色条纹的凹槽壁柱的菱形格子图案
Corner detail of decorative brickwork: diaper-patterned brickwork in contrasting colors with roundels and vertical stripes suggesting fluted pilasters

WA-199 多色英式交叉砌合的装饰窗下的带有拱心石的标准砖弧形拱窗
Gauged-brick segmental window arch with stone keystone under decorative brickwork of multicolored English cross bond

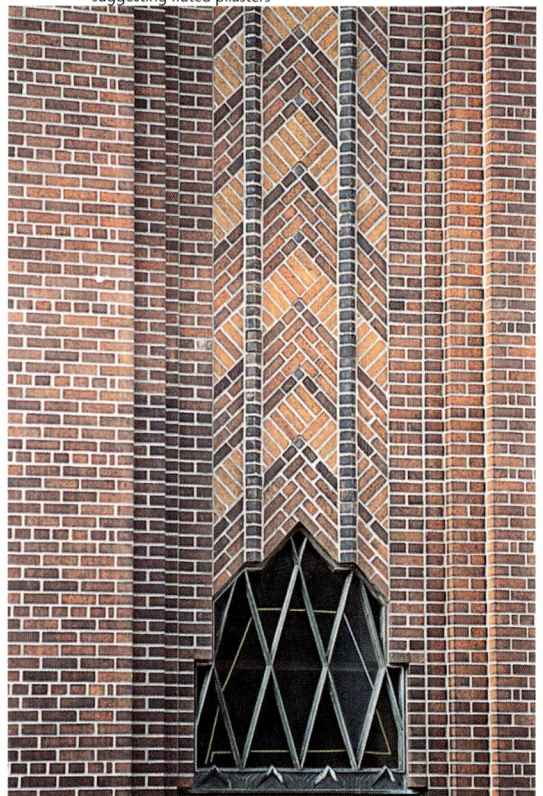

WA-200 顺砖砌合和曲线砌合中回纹图案的装饰砖结构，嵌壁板中的正门
Decorative brickwork in chevron pattern of running and stacked bonds; header rows framing grate in recessed pane

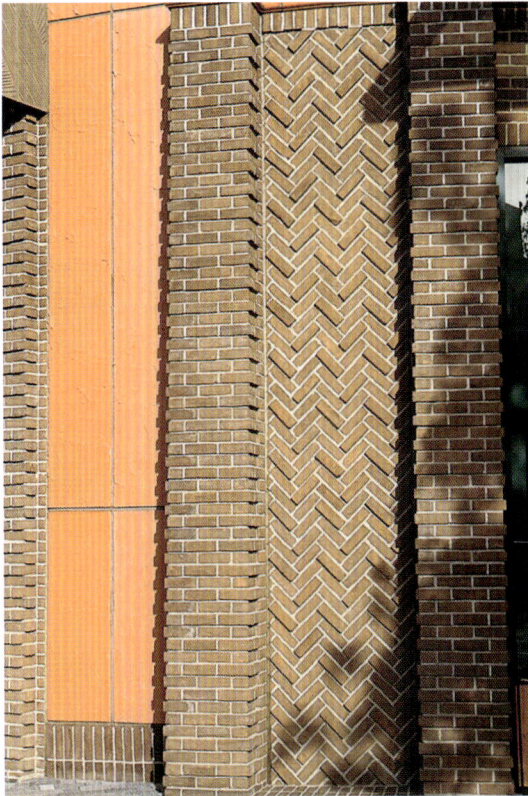

WA-201 顺砖砌合中带有多样齿边的壁柱；突出的人字形图案的装饰砖结构插板
Pilasters in running bond with variation of dog's-tooth edge; inset panel of decorative brickwork in herringbone pattern with projecting bricks

WA-202 装饰砖结构的隔带：突出立砌列的交替顺砖压缝砌合；立砖和立砌围绕着小圆孔，现代商业大厦
Bands of decorative brickwork: running bond alternating with projecting soldier rows; rowlock and soldiers around roundel, modern commercial building

WA-203 装饰砖带，从顶部到底部：回纹状水平砖带，菱形花格作品
Bands of decorative brickwork. Top to bottom: chevron, horizontal banding, diaper-work

WA-204 装饰砖带，从左到右：赤陶拱门，对缝顶梁，格子花纹，对缝顶梁，英式砌合
Bands of decorative brickwork. Left to right: terracotta arch, stacked header rows, basket-weave, stacked header rows, English bond

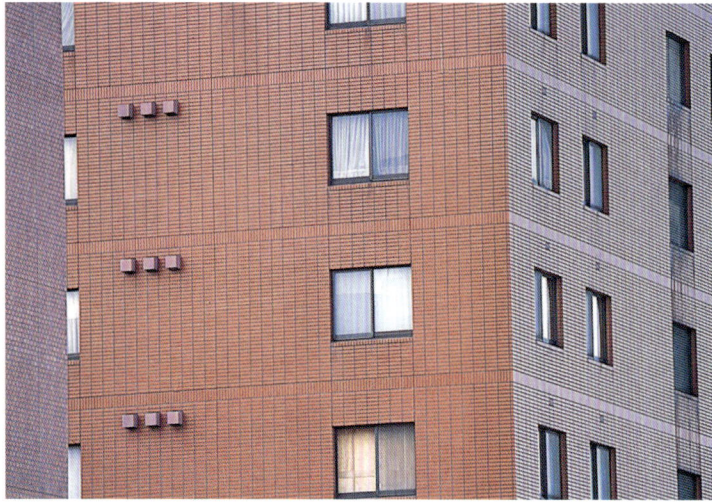

WA-205 对缝砌法中有精巧的竖砌砖带，现代公寓
Brick in stack bond with subtle soldier-course banding, modern apartment house

WA-206 不规则色调中带有突出斜线图案和方窗上沿的立砖砌合，现代商业大厦
Brick in random tones with projecting diagonal pattern and soldier courses above square windows, modern commercial building

WA-207 装饰砖结构，从左到右：方砖或带有砖制窗装饰的对角线图案的瓦片；横竖通缝砌法壁柱上的突出装饰，19世纪法国公寓楼
Decorative brickwork. Left to right: square brick or tile in diagonal pattern with brick window ornament; projecting ornament over stacked bond pilasters, 19th-c. French apartment house

砖墙面

WA-208 风化光面现浇混凝土，表面金属玻璃，现代房屋
Weathered cast-in-place concrete with smooth finish; metal and glass upper story, modern house

WA-209 采用网纹处理的木质模板浇筑的混凝土，住宅庭院
Board-formed concrete with textured finish from wood formwork, residential courtyard

WA-210 光面浇灌混凝土，连结在模板嵌板上，还有水平伸缩接头
Cast concrete with smooth finish, articulated formwork panels, and horizontal expansion joints

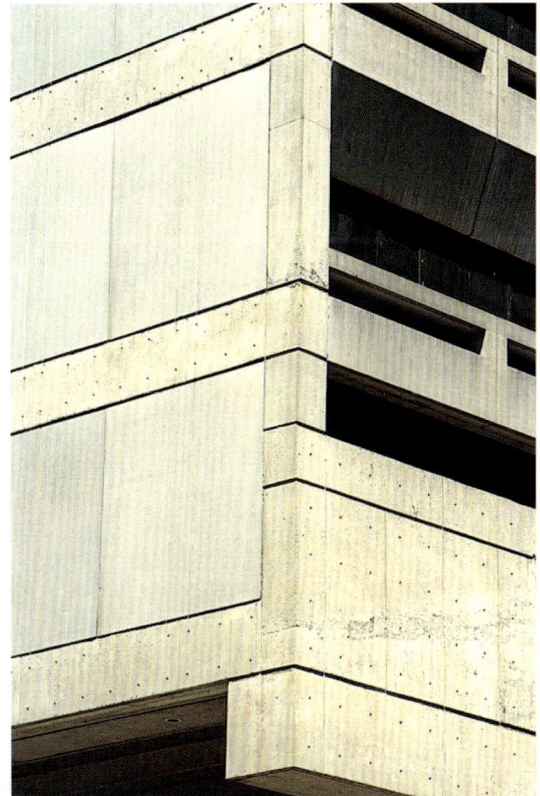

WA-211 带有砌合和连接嵌板的浇筑混凝土带的拐角细部，现代商业大厦
Corner detail of cast concrete with banding and interlocking panels, modern commercial building

灰泥和混凝土墙面

WA-212 彩色带状浇筑混凝土，现代商业大厦
Polychrome banded cast concrete, modern commercial building

WA-213 模板浇筑混凝土，带有木质网纹处理
Board-formed concrete with textured finish from wood formwork

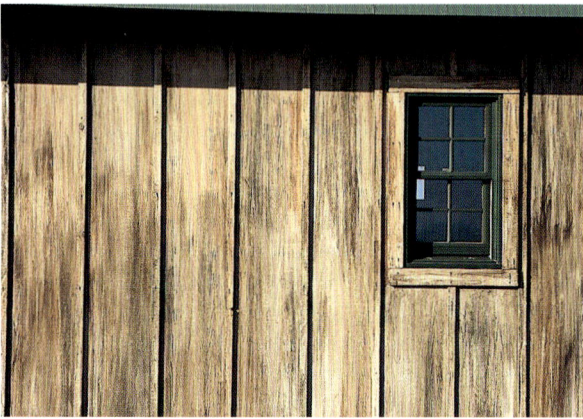

WA-214 有木板和板条图案铸成的混凝土板条嵌线
Concrete board siding cast with a wood board-and-batten pattern

WA-215 浇筑混凝土，浴室
Cast concrete, shower room

Photo: Austin, Patterson, Disston

WA-216 混凝土的突出窗口，现代房屋
Concrete projecting window, modern house

灰泥和混凝土墙面

WA-217 浇筑粗粒砂子与沥青混凝土，锤凿饰面，带厚边
Cast coarse-aggregate concrete, bush-hammered, with board-formed slab edge

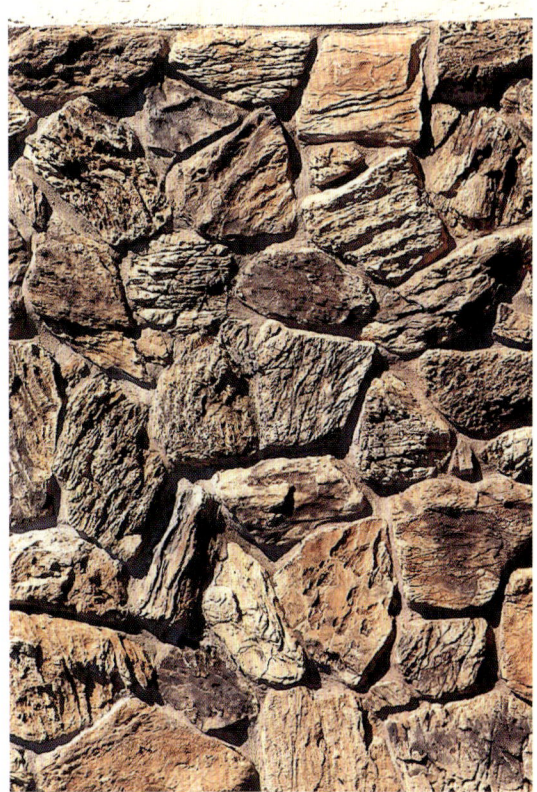

WA-218 碎石墙面中的人造石，住宅墙面
Manufactured stone in rubble wall, residential wall

WA-219 有回纹和菱形花格图案的预浇粗纹混凝土
Pre-cast textured concrete with chevron and diaper patterns

WA-220 预浇砂子与沥青混凝土，现代商业建筑
Pre-cast coarse-aggregate concrete, modern commercial building

灰泥和混凝土墙面

WA-221 不规则顺砖砌合中带有砂浆的风化水平混凝土大楼
Weathered handmade concrete block with mortar in irregular running bond on house

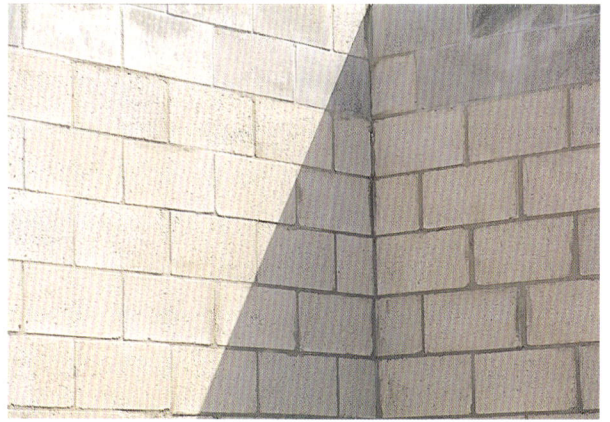

WA-222 顺砖砌合的砂浆框架墙的内角
Inside corner of mortared structural-block walls in running bond

WA-223 不规则顺砖砌合的砖混干墙，工业墙面
Concrete-block dry wall in irregular running bond, industrial wall

WA-224 模仿顺砖砌合中砖石建筑的拉毛水泥，用于矩形混凝土框架中
Stucco finish applied in rectangular concrete framework imitating masonry in running bond

WA-225 带有镶边和光面对比色砖的劈裂面混凝土墙的内外角
Interior and exterior corners of split-face concrete walls with banding and smooth-faced and contrasting color block, modern commercial building

灰泥和混凝土墙面

WA-226 带有粉色条带的顺砖砌合中的劈裂面混凝土建筑，现代商业大厦
Split-face concrete block in running bond with pink banding, modern commercial building

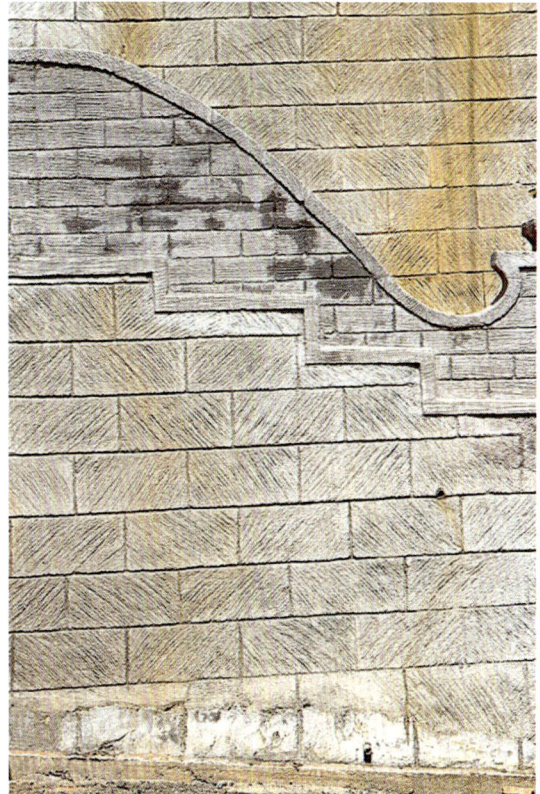

WA-227 顺砖砌合中，雕刻防护墙下带有梯状带的雕刻混凝土建筑，网纹处理的矩形石
Carved concrete block textured to resemble sawn stone in running bond with stepped band under curved guard wall

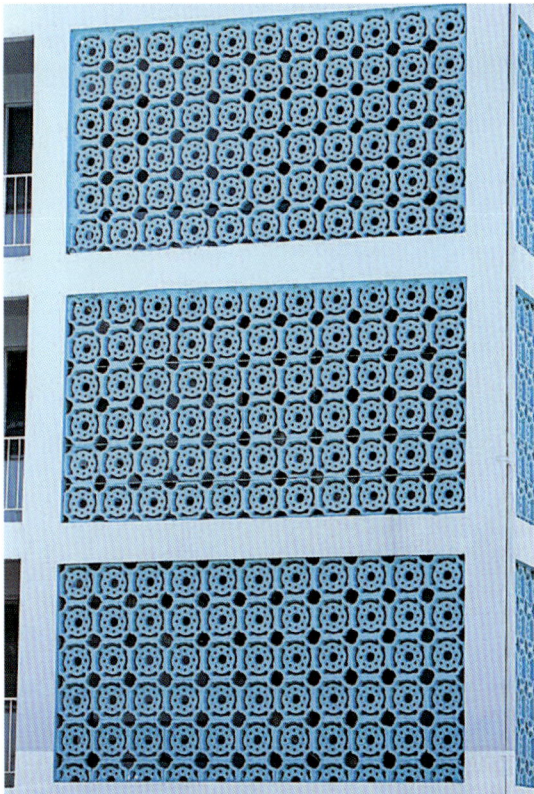

WA-228 屏挡墙的定制混凝土建筑，住宅门廊
Custom concrete blocks in screen wall, residential entryway

WA-229 带有屏挡墙（顶部）的定制混凝土部件，住宅门廊
Custom concrete blocks with screen wall (top), residential entryway

灰泥和混凝土墙面

WA-230 蜿蜒图案设计的拉毛水泥面、石块镶边窗口和角落里的楔形石以及天然石头的嵌入；百叶窗，房屋
Stucco with dimension-stone window and corner quoins and natural stone insets in serpentine design; louvered shutters, house

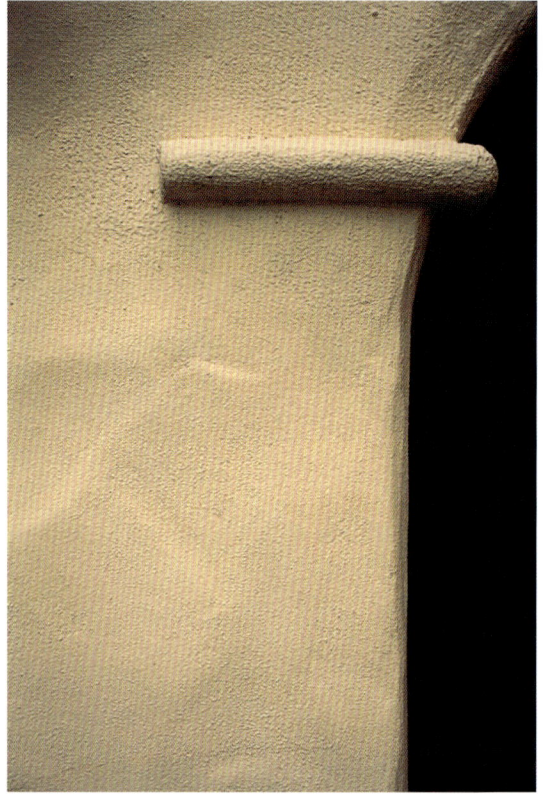

WA-231 配有拱形线脚的拱形出口细部，砖坯上附有拉毛水泥，美国西部建筑
Detail of arched opening with impost-like molding, stucco over adobe, American Southwestern mission

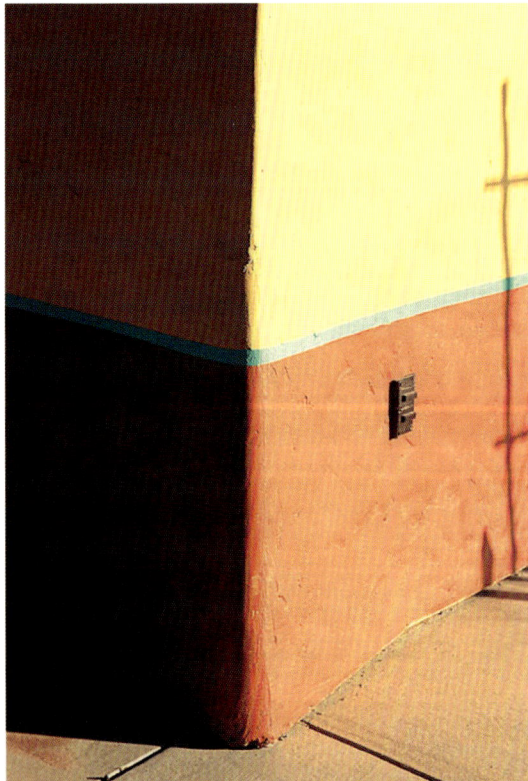

WA-232 底部逐渐加宽的圆滑拐角，模仿美国西部土墙房屋细部的光面喷绘拉毛水泥拐角
Flared, rounded corner of smoothed painted stucco imitating a detail of traditional American Southwestern adobe house

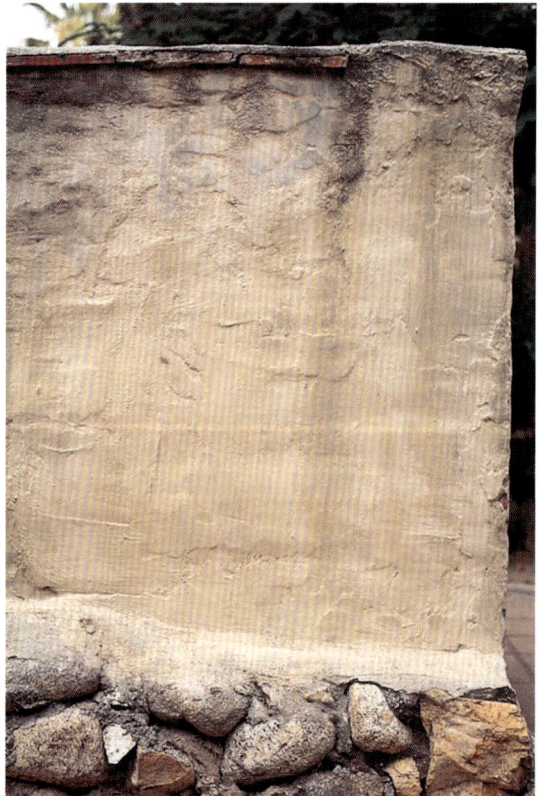

WA-233 粗石基座上附于砖坯面上的光面石灰泥墙，庭院墙
Smoothed and whitewashed mud plaster over adobe brick on a base of fieldstone, courtyard wall

灰泥和混凝土墙面

WA-234 网纹拉毛水泥建筑的光面水泥楔形石和带状隔带
Smooth stucco quoins and belt course on textured stucco building

WA-235 山墙的波浪形拉毛水泥面，工业建筑
Undulating stucco finish on gable-end wall, industrial building

WA-236 配有光面水泥带状突出部分的彩色粗糙抹泥面，意大利别墅风格的现代商业大厦
Colored rough-troweled finish on stucco with smooth stucco band forming jetty, Italian-villa-style modern commercial building

WA-237 沿着突出部分的薄浆砖缝内接设计，现代商业大厦
Inscribed design in grouted blocks running along jetty, modern commercial building

灰泥和混凝土墙面

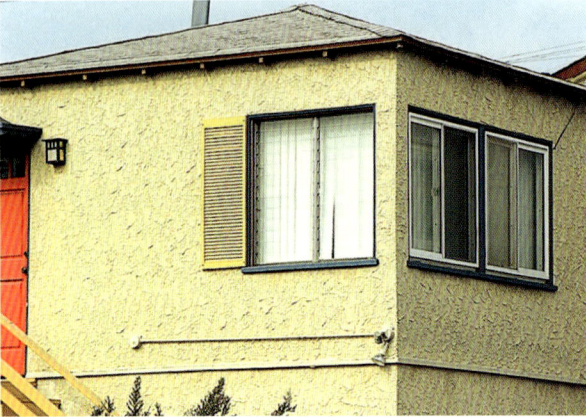

WA-238 海滩别墅的纹理抹面拉毛水泥面
Knocked-down skip-troweled stucco finish on beach house

WA-239 网纹和内雕刻拉毛水泥，加叶饰的和有图片的设计同顺砖砌合中的分割面石的仿造物混合在一起
Textured and inscribed stucco, foliated and pictorial designs mixed with imitation split-face stone in running bond

WA-240 浇筑混凝土对缝砌合，以儿童手印作装饰的基座，城市公共空间
Cast concrete walls imitating stack bond, base course decorated with imprints of children's hands, urban public space

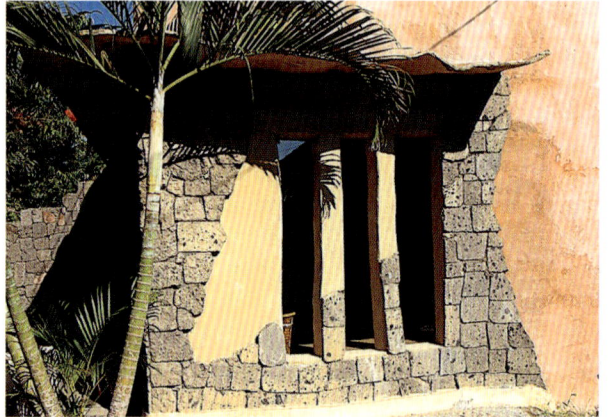

WA-241 随意砖石结构石头表面的拉毛水泥
Stucco over random ashlar stone

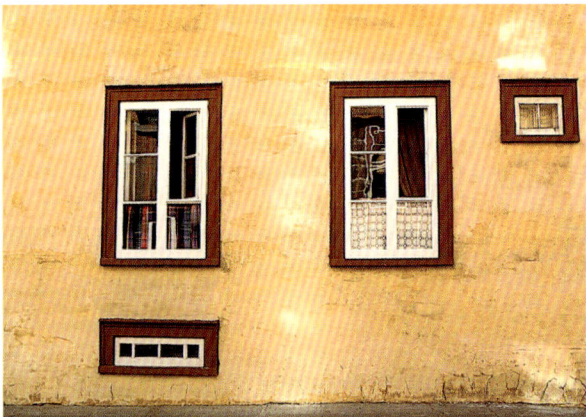

WA-242 带有不同窗户的喷漆墙面的风化波浪形光滑水泥面，城市房屋
Weathered undulating smooth stucco on painted wall with a variety of windows, urban house

WA-243 带有彩色水泥层的泥墙，庭院墙壁
Mud wall with layers of tinted mud stucco, courtyard wall

灰泥和混凝土墙面

WA-244 有百叶窗的石灰泥波浪状的光面水泥墙，公寓
Whitewashed undulating smooth stucco wall with shuttered windows, apartment house

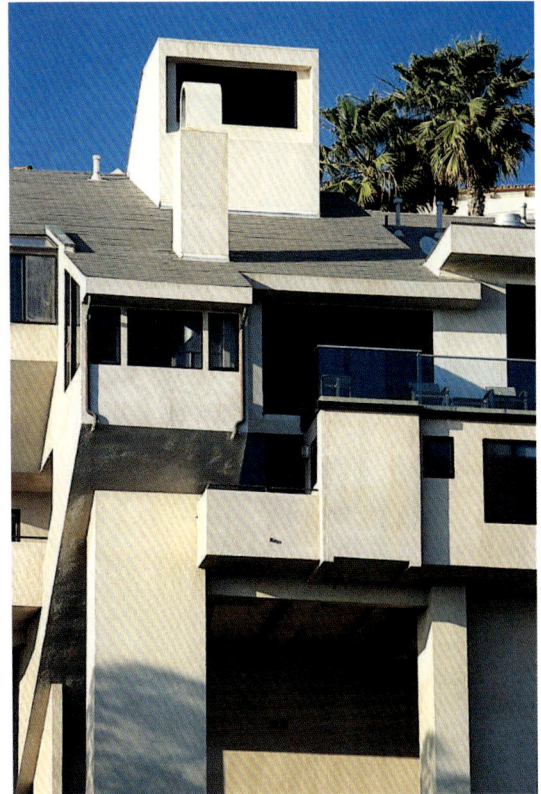

WA-245 喷漆光面拉毛水泥，现代房屋
Painted smooth stucco, modern house

WA-246 墙上带有磨面彩色水泥；双挂窗，意大利别墅风格现代公寓
Colored stucco with knocked-down skip-trowel finish on wall; double hung windows, Italian-villa-style modern apartment house

WA-247 低层部分以水泥楔形石作为图案的彩色光面水泥墙，现代公寓
Colored smooth stucco supergraphic with stucco quoins in lower section, modern apartment house

灰泥和混凝土墙面

WA-248 风化彩色光面拉毛水泥，壁柱和拱窗
Weathered colored smooth stucco; pilasters and arched windows

WA-249 风化山墙上的有中砾石和楔形石的大骨料拉毛水泥面
Large-aggregate stucco with cobbles and quoins on weathered gable-end wall, farm building

WA-250 带格子窗和装饰框的屋脊盖瓦，风化彩色光滑拉毛水泥环绕椭圆形窗
Weathered colored smooth stucco surrounding oval window with grille and ornate frame; hipped tile roof

灰泥和混凝土墙面

摄影: Meredith Barchat

WA-251 山墙上的彩色拉毛水泥网纹
Painted stucco textures on gable-end wall

摄影: Meredith Barchat

WA-252 风化拉毛水泥墙环绕着带水泥墩的露台
Weathered stucco wall surrounding balcony with stucco-covered pillars

WA-253 风化的彩色拉毛水泥裸露的冲压泥墙
Weathered tinted mud stucco revealing rammed mud wall

灰泥和混凝土墙面

WA-254 风化喷漆灰泥显露结构石和木包层
Weathered painted plaster revealing structural stone and wood cladding

WA-255 带露台的庭院墙上的风化彩色光面拉毛水泥
Weathered colored smooth stucco on courtyard wall with balcony

WA-256 风化彩色拉毛水泥，木质过梁裸露在外，打底灰泥层，结构砖石，罗马式砖
Weathered colored stucco revealing timber lintel, scratch coat and structural stone, and Roman brick

WA-257 风化彩色拉毛水泥，意大利式建筑
Weathered colored stucco, Italianate building

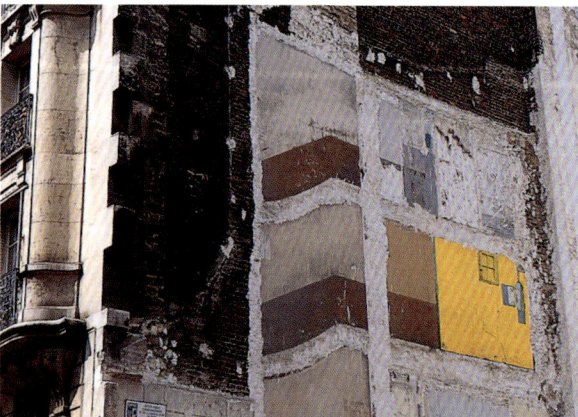

WA-258 由于拆毁而裸露在外的灰泥内部墙面
Walls of plastered interior apartment walls exposed by demolition

WA-259 裸露住宅庭院砖墙外层的风化石灰泥，美国西南部地区
Weathered whitewashed mud-stucco over exposed adobe brick courtyard wall, American Southwestern mission

灰泥和混凝土墙面

WA-260 风化拉毛水泥的裸露结构砖石，新颖的灰泥建筑，包括人造楔形石和顺砖砌合的砖块
Weathered stucco exposing structural brick and stone. Original plasterwork included faux-stone quoins and blocks in running bond

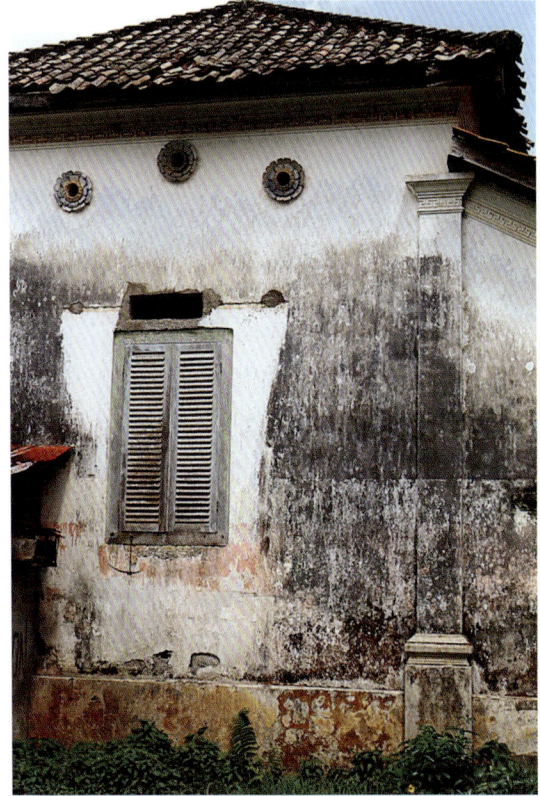

WA-261 带壁柱的风化拉毛水泥墙
Weathered stucco wall with pilaster

WA-262 花园墙面上有热带植被的风化白色拉毛水泥墙
Weathered whitewashed mud-stucco withinvasive tropical vegetation on garden wall

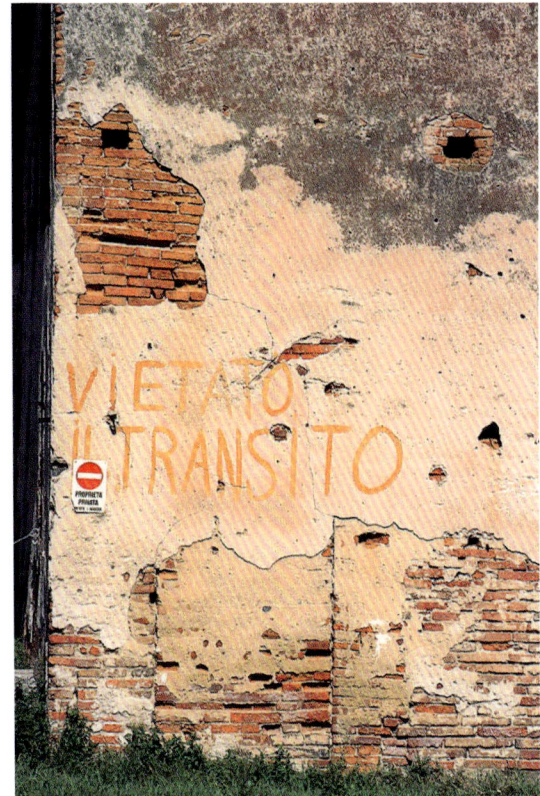

WA-263 有混凝土斑纹、涂鸦和裸露结构砖石的风化拉毛水泥墙
Weathered stucco with concrete patches,graffiti, and exposed structural brick and stone

灰泥和混凝土墙面

WA-264 庭院墙，从顶部到底部：立砖压顶和顶梁砖遮檐；铸铁混凝土（左）；顺砖砌合砖柱；方形粗石基座
Courtyard wall. Top to bottom: rowlock and header brick coping; cast concrete (left); running-bond brick pilaster; roughly squared fieldstone base

WA-265 私人车道入口处的砖和粗石的街灯柱，艺术和工艺品住宅
Brick-and-fieldstone lamppost at driveway entry, Arts and Crafts residence

WA-266 庭院墙，从顶部到底部：浇筑混凝土砖墩；锯齿砖带；浅色石，混凝土框架中的人字形图案砖；彩喷砖；石头基座
Courtyard wall. Top to bottom: cast-concrete coping; saw-tooth brick course; light-colored stone, herringbone-bond brick in concrete frame; painted brick; stone base

WA-267 砖框架层列琢石墙面结构嵌板，花园墙
Coursed ashlar stone panel in brick frame, garden wall

WA-268 单层板岩同双层砖交替的顺砖砌合，花园墙
Single courses of slate alternating with double courses of brick in running bond, garden wall

WA-269 带有不规则形状粗方砖的混砌砖
Random-bond brick with roughly squared fieldstone

混合材料

WA-270 风化杉木垂直木板，适合粗石基座的侧面形状
Vertical planks of weathered cedar shaped to fit profile of fieldstone base

WA-271 带有化石遮檐和基座的彩色拉毛水泥墙
Colored stucco wall with fossil-stone coping and base

WA-272 巨砾和大鹅卵石随意地同定制砌合砖布置在一起，开口被铁栏或幕板填
满，破损的花园墙壁
Boulders and large cobbles randomly interspersed with custom-bond brick, openings filled
with grillework or screen blocks, battered garden wall

混合材料

WA-273 半砖木结构和固定结构上的菱形嵌板
Diamond-shaped panels on half-timbered and pegged construction

WA-274 拉毛水泥和厚木填实；木框建筑；传统日式木格子窗
Stucco and wood-planked infill; timber-frame construction; traditional Japanese wood lattice window

WA-275 拉毛水泥填实；在梁柱交叉处有金属装饰的梁柱墙，并带有钟形的滑动闸板
Stucco infill; post-and-beam wall with decorative metal tie-irons at intersections of posts and beams, bell-shaped slidingshutter

WA-276 房屋突出部分下有梁托的半木建筑；宽大的窗户和飘窗下的伸臂支柱，16世纪或17世纪英国房屋
Half-timber construction with corbels under jetty; leaded windows, and console supports under oriel, 16th- or 17th-c. English house

WA-277 有砖填实的半木结构（顶部）和拉毛水泥（底部），17世纪法国房屋
Half-timbering with brick infill (top) and stucco (bottom), 17th-c. French house

混合材料

WA-278 有水泥填充和铜色"木料"的装饰性半木外立面，20世纪住宅建筑
Decorative half-timbered facade with concrete infill and copper timbers, 20th-c. residential building

WA-279 拉毛水泥和厚木板填充的梁柱结构，部分装饰木柱，传统日式建筑
Post-and-beam construction with stucco and wood-plank infill; partially dressed timber post, traditional Japanese building

WA-280 半木结构；房屋突出部分下的粗石，16世纪或17世纪英国房屋
Half-timbering; fieldstone under jetty, 16th- or 17th-c. English house

WA-281 波浪形拉毛水泥和厚框架般的木制框架，传统日式建筑
Timber frame with infill of undulating stucco and framed planks, traditional Japanese building

混合材料

WA-282 砖填半木结构；一层为粗石，16世纪或17世纪英国房屋
Half-timber with brick infill; fieldstone first story, 16th- or 17th-c. English house

WA-283 拉毛水泥填充的半木结构，15世纪或16世纪法国房屋
Half-timber with stucco infill, 15th- or 16th-c. French house

WA-284 砖填的半木结构，15世纪或16世纪法国房屋
Half-timber with brick infill, 15th- or 16th-c. French house

WA-285 带有平木"木材"和砖填的装饰性半木外立面和凸窗，20世纪房屋
Decorative half-timbered facade with flat wood "timbers", brick infill, and bay windows, 20th-c. house

混合材料

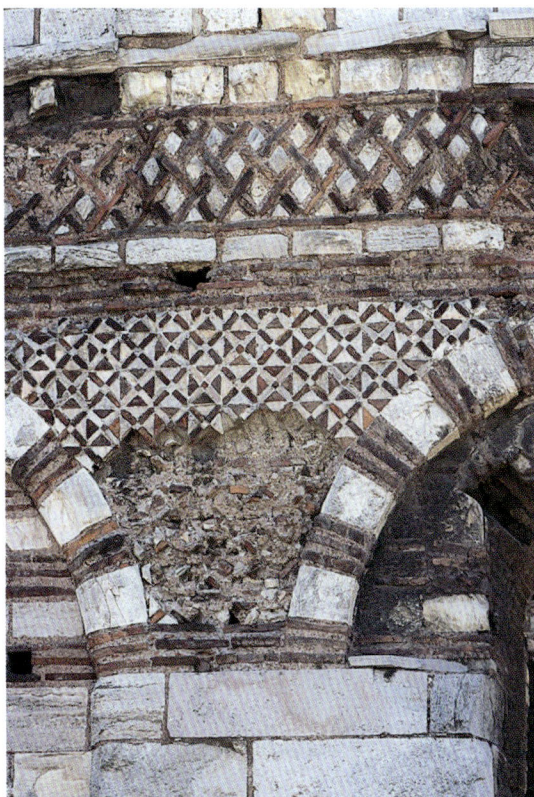

WA-286 古老的商店，从顶部到底部：不规则的方形粗石；
菱形花格砖结构；石带和罗马砖层，有下部砖石拱
的风化瓷砖；层列琢石墙面石
Ancient arcade. Top to bottom: roughly squared stone; diaper
brickwork; courses of stone and Roman brick, weathered tile
with substructure, brick-and-stone arches; coursed-ashlar stone

WA-287 拱心石和供腹上有雕刻装饰的带状砖石拱
Banded brick-and-stone arches with carved ornament in
keystones and spandrels

WA-288 有砖框窗和砖带层列的非层列碎石墙；有格子窗和
百叶窗的突起拱窗
Uncoursed rubble wall with brick-framed windows and brick belt
courses; cambered arch window with trompe l'oeil grille and
shutters

WA-289 带梁托出挑和邻近的上一层砖墙
Coursed ashlar on inside-corner turret with corbelled jetty and
adjacent upper story; remaining walls of brick

混合材料

WA-290 顺砖砌合中的粗面石墩，由砖环绕着；卵锚饰壁柱冠和前线的两层之间配有浮雕装饰带
Rusticated stone blocks in running bond, surrounded by brick; egg-and-dart pilaster cap and molding, with relief ornament band between stories

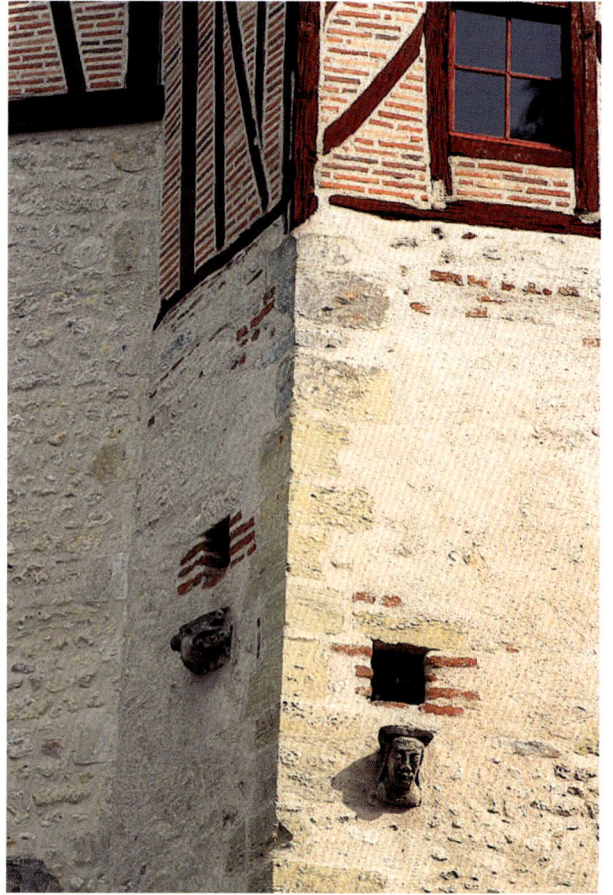

WA-291 半砖木结构的上半部和有楔形石的拉毛水泥下半部
Half-timber upper story with brick infill and stucco lower story with stone quoins

WA-292 有几何图形的上下层窗间墙面的砖和楔形石
Brick with stone quoins between spandrels with geometric tile patterns

混合材料

WA-293 黑色花岗岩基座配有金属玻璃幕墙，现代商业大厦
Black granite base with metal-and-glass curtain wall, modern commercial building

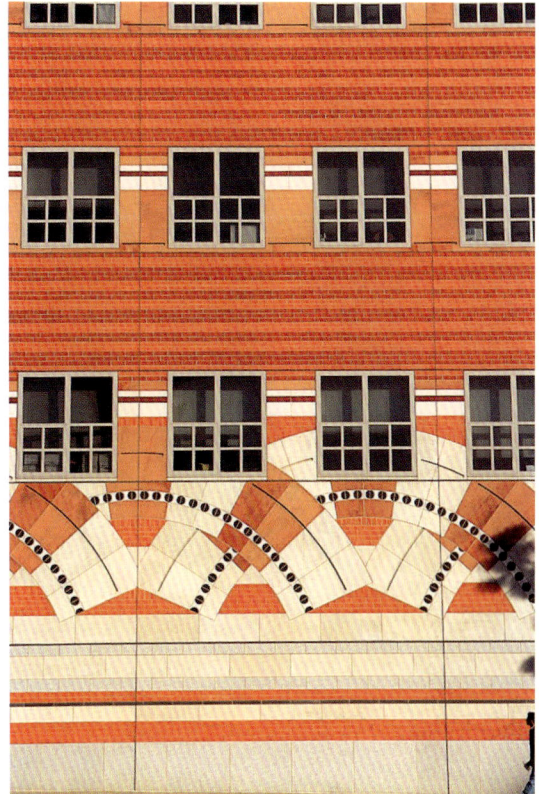

WA-294 浇筑混凝土或石料的上半部有大图案的重复带状顺砖砌合砖石结构，现代商业大厦
Banded running-bond brickwork on upper stories with supergraphic running repeat in cast concrete or stone, modern commercial building

WA-295 由石头或预浇混凝土顺砖砌合镶边的砖
Brick in running-bond bands with stone or pre-cast concrete, modern commercial building

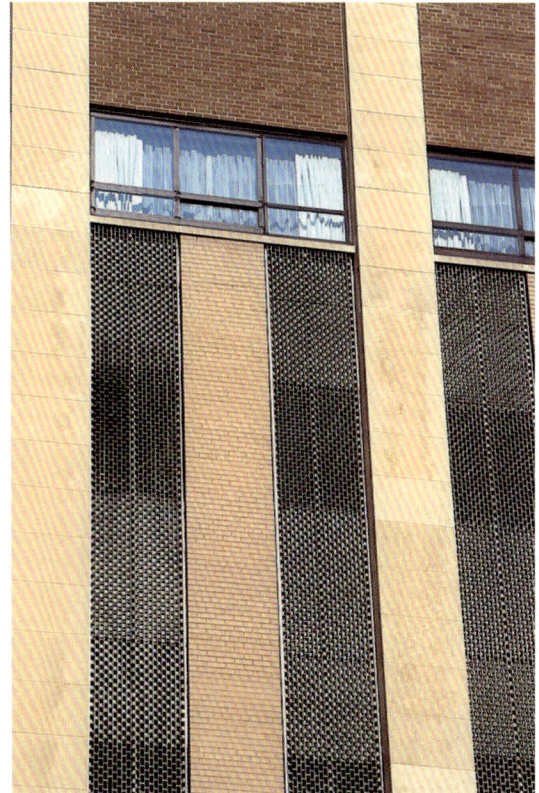

WA-296 砖、金属网、石头，现代商业大厦
Brick, metal mesh, and stone, modern commercial building

混合材料

WA-297 不锈钢电镀和砖，现代商业大厦
Stainless steel cladding and brick, modern commercial building

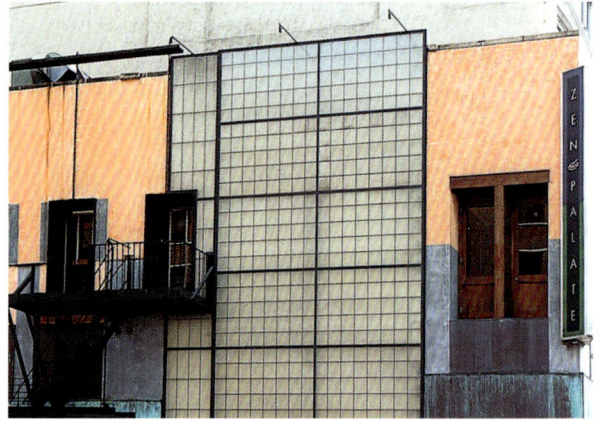

WA-298 有彩色和生了铜绿的金属框玻璃嵌板，现代商业大厦
Metal-framed glass panels with colored and patinated metal, modern commercial building

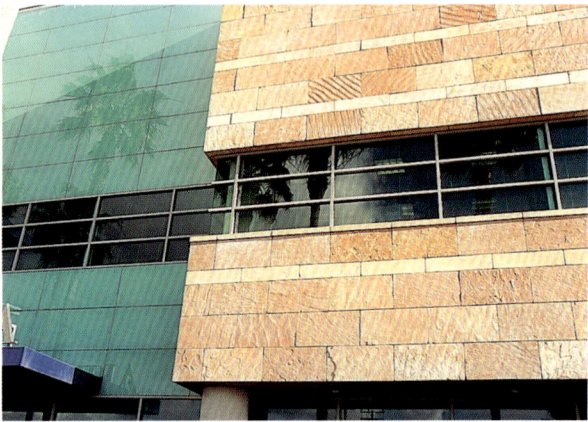

WA-299 开裂砂石贴面板，旁边是彩色玻璃或塑胶玻璃贴面，现代商业大厦
Split-face sandstone veneer next to colored-glass or plexiglass-clad wing, modern commercial building

WA-300 有玻璃砖和瓷砖（左上）的劈裂面砂石，现代商业大厦
Split-face sandstone with glass block and tile (top left), modern commercial building

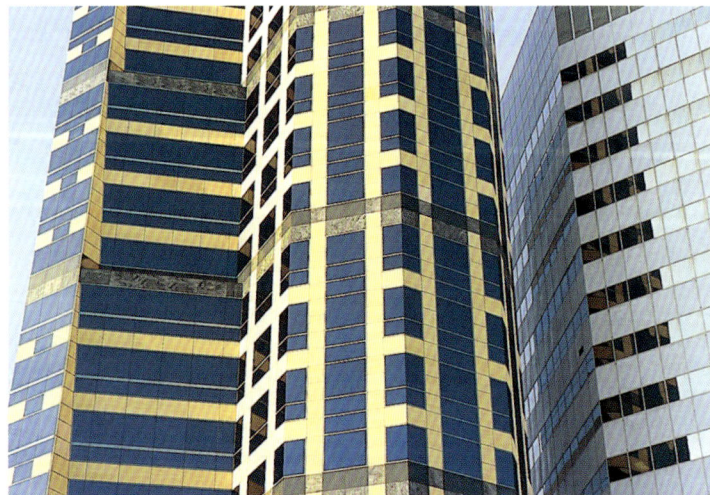

WA-301 玻璃和石头幕墙，现代商业大厦
Glass and stone curtain walls, modern commercial building

混合材料

WA-302 从顶部到底部：金属面板板条；浇筑混凝土或石料；金属卵石；金属框架玻璃，现代商业大厦
Top to bottom: batten-seam metal cladding; cast concrete or stone; metal shingles; metal-framed glass, modern commercial building

WA-303 带有金属边饰的混凝土建筑
Concrete building with metal trim

WA-304 石头贴面板和有贴面楔形石的砖翼，现代商业大厦
Stone veneer and brick wing with stone-veneer quoins, modern commercial building

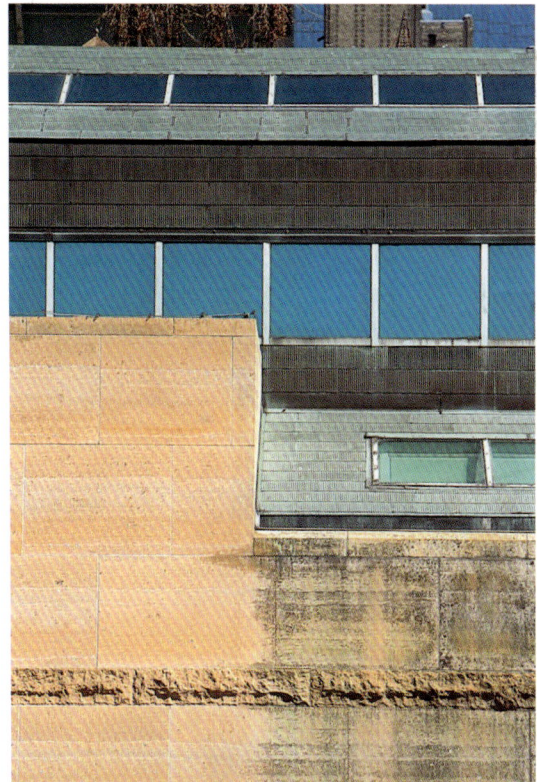

WA-305 从顶部到底部：波浪状铜绿金属面板；层列琢石墙面结构，现代商业大厦
Top to bottom: patinated corrugated metal cladding; coursed ashlar sandstone with rock-face belt course, modern commercial building

混合材料

WA-306 底部有梯状浇筑混凝土的玻璃砖塔楼，旁边是混凝土的装饰艺术风格现代住宅
Glass-block tower with stepped cast concrete at base next to cast-concrete Art Deco–style modern residential building

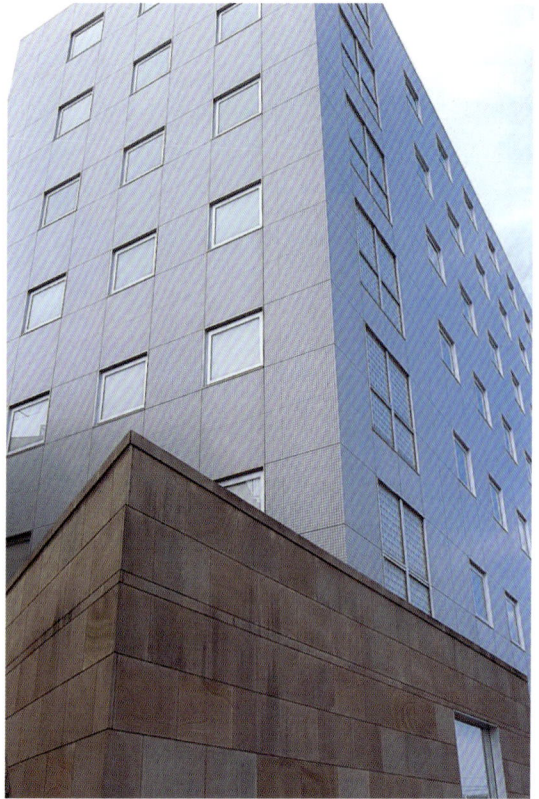

WA-307 石头贴面底层和包瓷砖上层，现代商业大厦
Stone-veneer lower story and tile-clad upper stories, modern commercial building

WA-308 玻璃、石头和砖构造的现代商业大厦
Glass, stone, and brick modern commercial building

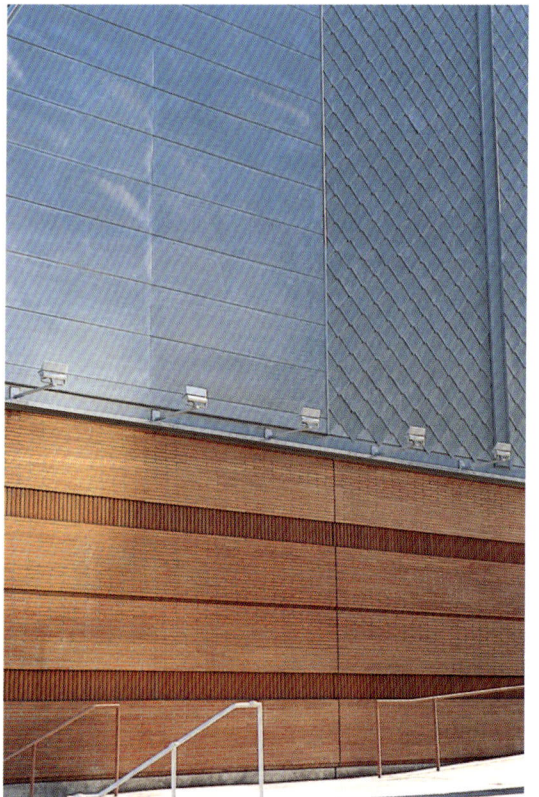

WA-309 下部有砖带的内部金属面板，现代工业建筑
Interior metal cladding with banded brick below, modern industrial building

混合材料

外立面

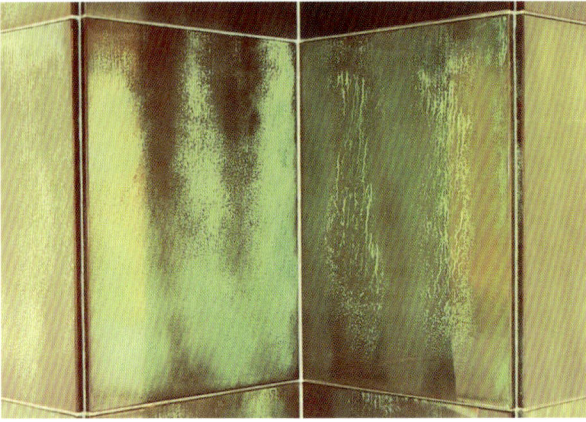

FA-1　净胶铜绿面板，灌浆接缝细部
Corner detail of clear-coated verdigris cladding, with grouted seams

FA-2　钢或铝墙立接缝面板，现代商业大厦
Steel or aluminum wall standing-seam cladding, modern commercial building

FA-3　风化的压制成形的锡中插入人造石头
Weathered stamped-tin sheathing in faux stone pattern

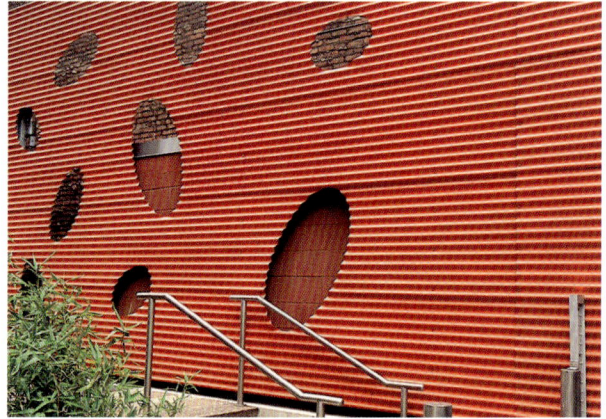

FA-4　有椭圆镂空图案的彩喷波浪金属幕墙，城市公园
Painted corrugated metal screen wall with elliptical cutouts, urban park

FA-5　上下层窗间的不锈钢嵌板外的青铜（右上）
Bronze over stainless steel spandrel panel with a raised edge and adjoining window (top right)

FA-6　配有扣拴的风化平板金属嵌板，可能是搭接的，工业建筑外墙
Weathered flat metal panels with visible fasteners, probably lapped, exterior wall of industrial building

金属和玻璃

FA-7 金属面现代房屋，顶部：近似垂直的板条；毛条窗；
底部：齐平嵌板（左）；对比色的金属入口（右）
Metal-clad modern house. Top: adjacent vertical strips; strip
window; bottom: flush panels (left); contrasting metal entry (right)

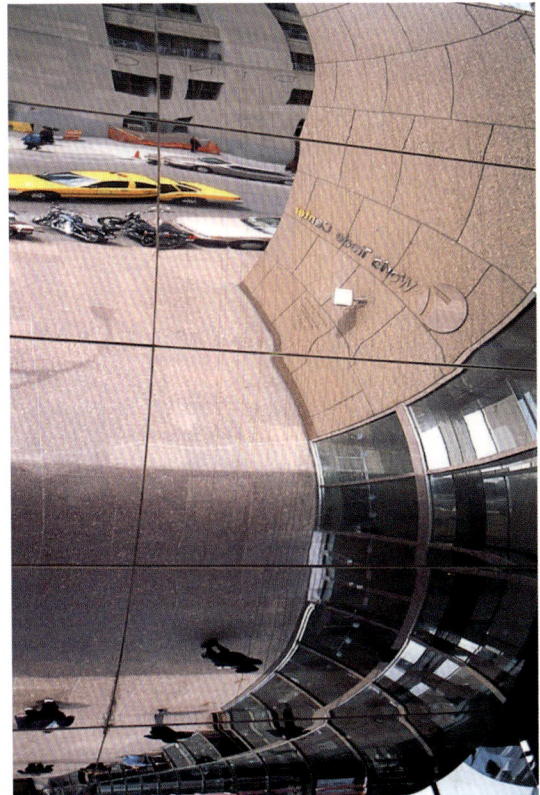

FA-8 齐平弯曲镜面不锈钢面板，现代商业大厦
Flush curved mirror-finished stainless steel cladding, modern
commercial building

FA-9 有压制图案装饰线脚和裸露扣拴的刷面拉毛不锈钢面
板，美国餐厅
Brushed stainless steel cladding with stamped ornamental moldings
and exposed fasteners, American diner

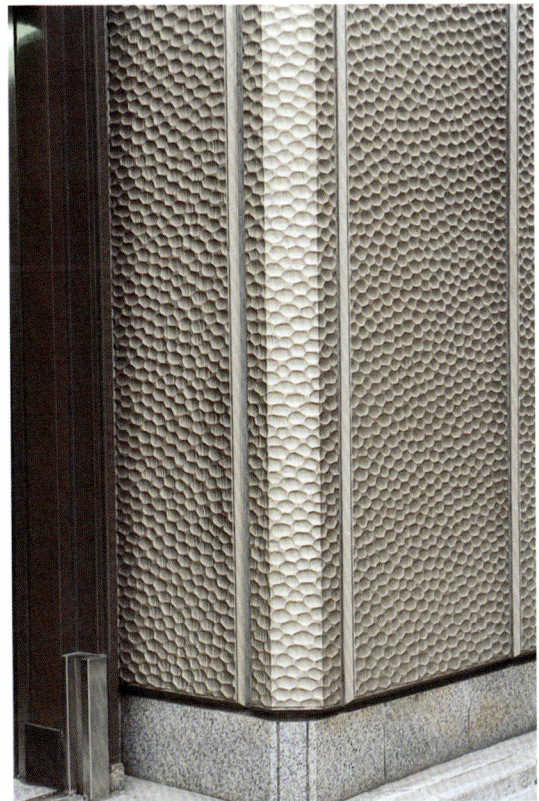

FA-10 带有凿面的垂直面板
Vertical cladding with hammered finish

金属和玻璃

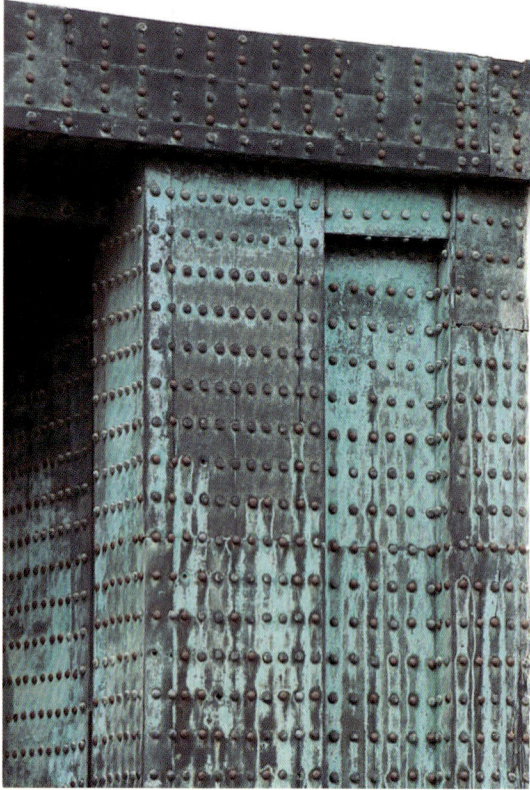

FA-11　突出扣拴的风化和自然铜绿桥墩，古老的日式入口
Weathered and naturally patinated pier with prominent fasteners,
ancient Japanese entry

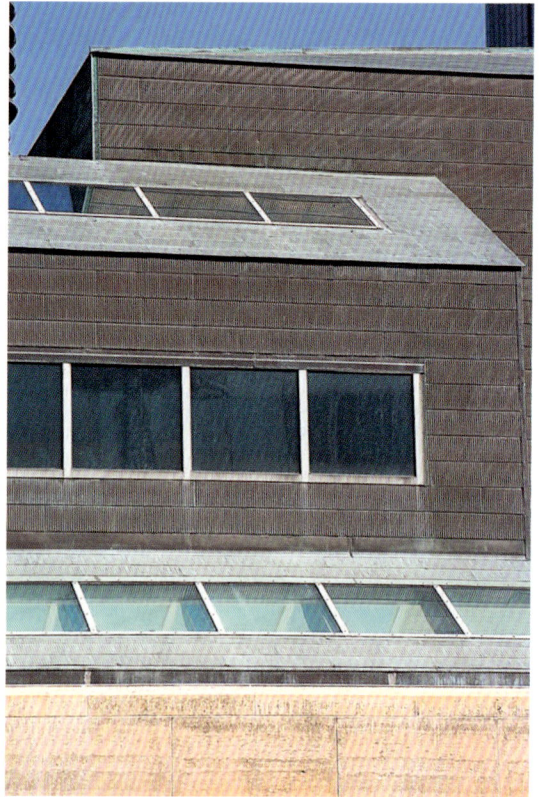

FA-12　波浪钢铜绿盖板围绕着毛条窗和天窗
Corrugated metal patinated shingles surrounding strip windows
and skylights

FA-13　镀铜覆面，现代商业大厦
Copper cladding, modern commercial building

FA-14　幕墙；有几何立体图形的上下层窗间墙面嵌板，现代
商业大厦
Curtain wall; spandrel panels with dimensional geometric pattern,
modern commercial building

金 属 和 玻 璃

FA-15　铸钢板，现代工业建筑入口
Flush cast-bronze cladding, entry of modern institutional building

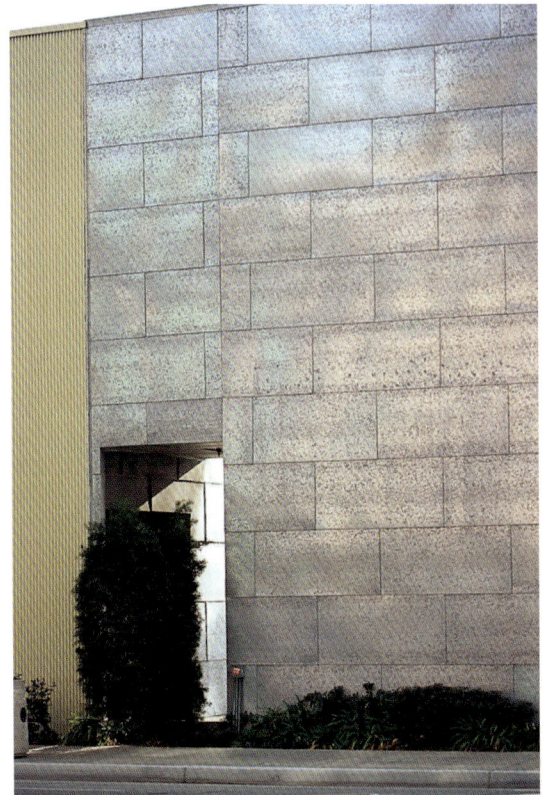

FA-16　顺砖砌合图案中的镀锌金属面板
Galvanized metal cladding in running-bond pattern

FA-17　铜嵌板，带有垂直接缝，水平搭接
Copper panels, possibly with standing vertical seams, lapped horizontally

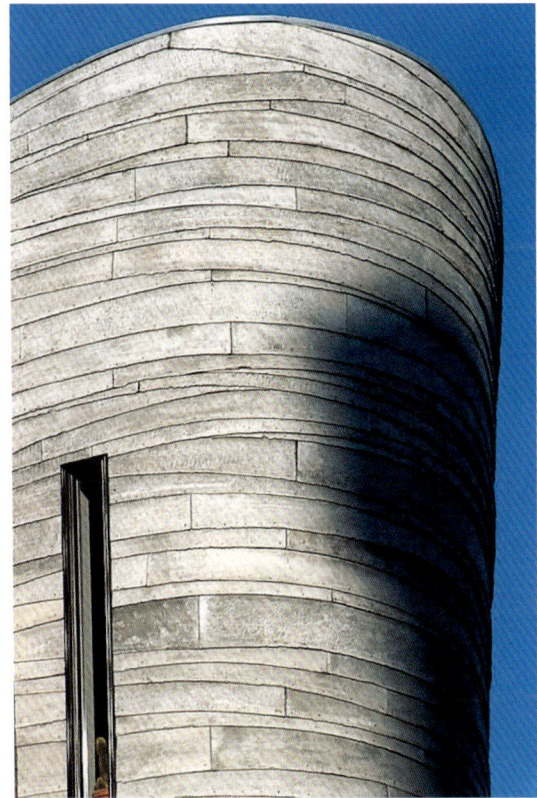

FA-18　金属条包裹着带有不规则板条和露明固件的半圆外立面
Metal strips wrapped around a semicircular facade with irregular laps and visible fasteners

金属和玻璃

FA-19　镀锌钢，板条层垂直嵌板
Possibly galvanized steel, batten-seam vertical panels

FA-20　有露明固件的波浪形镀锌钢板，现代商业大厦
Corrugated galvanized steel panels with visible fasteners, modern commercial building

FA-21　有露明固件和毛条窗的喷漆浮雕玻璃钢板，现代住宅建筑
Painted curved corrugated metal panels with visible fasteners and strip windows, modern residential building

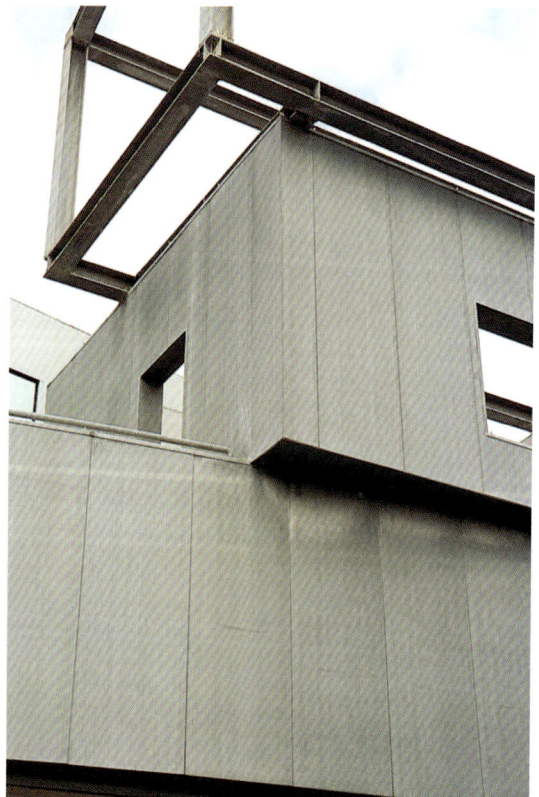

FA-22　平贴的镀锌板，现代商业大厦
Flush galvanized panels, modern commercial building

金属和玻璃

FA-23   有冲压菱形图案的风化镶嵌墙体系，现代公寓
Weathered panelized wall system with stamped diamond pattern, modern apartment house

FA-24   搭缝彩喷和铜绿金属板，现代商业大厦
Painted and patinated metal panels with lapped seams, modern commercial building

FA-25   有黑色酸蚀不锈钢嵌板和窗户的加强混凝土框架，现代公共机构建筑
Reinforced concrete frame with black acid-etched stainless steel panels and windows,
modern institutional building

金 属 和 玻 璃

FA-26 有搭缝的不锈钢小块挡板，现代公共机构建筑
Stainless steel shingle cladding with lapped seams, modern institutional
building

FA-27 定制装配金属板外立面上的卵形凸出，现代商业大厦
Ovoid projection on facade made of custom fabricated metal panels,
modern commercial building

FA-28 空间架上带三角图案的印花金属板，现代商业大厦入口
Stamped metal panels in a triangular pattern on space frame, entry to modern commercial building

金属和玻璃

FA-29 有镜面玻璃的全玻璃弯曲幕墙
Fully glazed curved curtain walls with mirror glass

FA-30 带状幕墙，镜面玻璃和上下层窗间实心板交替出现
Banded curtain wall, alternating mirror glass and solid spandrel panels

摄影: Guy Gurney

FA-31 带状幕墙，上下层窗间镜面玻璃和窗户交替出现
Banded curtain wall, alternating mirror glass spandrels and windows

幕墙

FA-32　交错的上下层窗间墙元素构成的外墙板系统，类似预浇铸混凝土和玻璃
Cladding system of staggered spandrel elements, probably precast concrete and glass

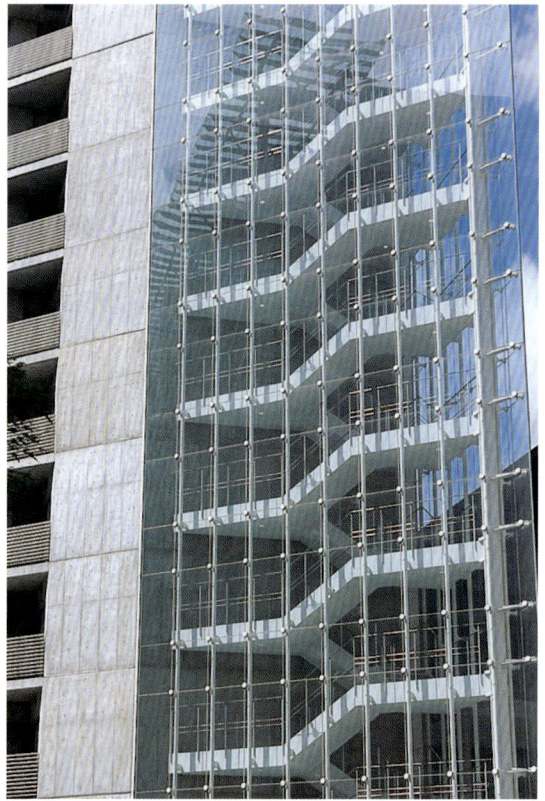

FA-33　幕墙围绕着有玻璃板的楼梯墙，通过托臂系统和扣拴固定在拐角上，用硅树脂密封
Curtain wall around a stair tower with glass panels secured at corners by a bracket system and seams sealed with silicone

FA-34　带状幕墙，镜面玻璃和固定的上下层窗间墙面板交替排列
Banded curtain wall, alternating mirror glass and solid spandrel panels

FA-35　有圆形拐角的玻璃砖幕墙
Glass-block curtain wall with rounded corner

幕墙

FA-36 从左到右：全玻璃表面；固定的上下层窗间墙面板和玻璃交替排列；带柱面装饰和玻璃部件的幕墙系统
Left to right: fully glazed; alternating solid spandrel panels and glazing; curtain-wall system with column covers and glazed units

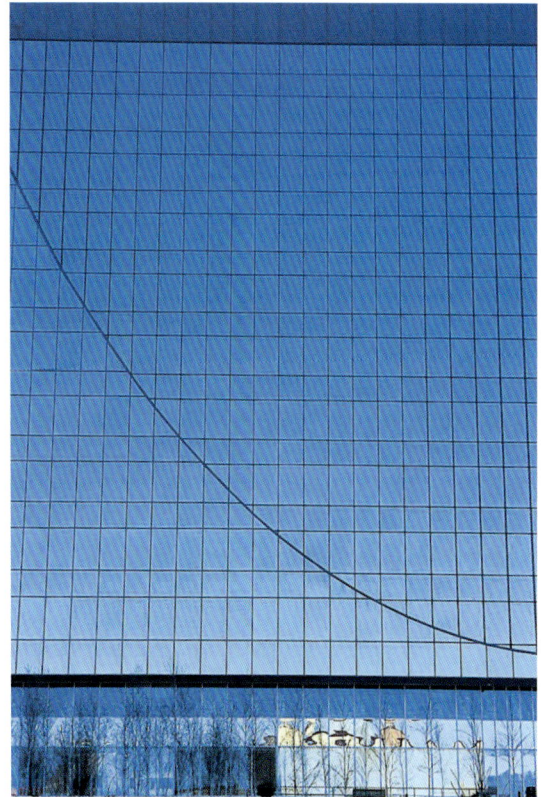

FA-37 全玻璃幕墙；底部嵌壁式镜面玻璃反射出庭院的树木
Fully glazed curtain wall with glass inset above curve; recessed mirror glass at bottom reflects trees in courtyard

FA-38 全玻璃幕墙，百叶窗在底层；突出物类似托架系统或装饰
Fully glazed curtain wall, shading louvers visible at lower level. Projections may be structural bracket system or ornamental

FA-39 全玻璃幕墙的嵌入式覆盖系统，玻璃幕墙带有浅色光谱效果玻璃或塑料封闭电梯井
Panelized cladding system with fully glazed curtain wall of light spectrum-producing glass or plastic enclosing elevator shaft

幕墙

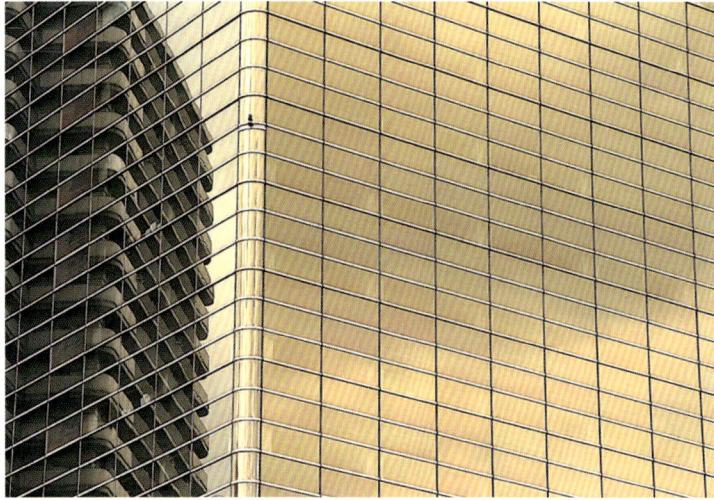

FA-40 有水平竖框的茶色镜面玻璃幕墙
Fully glazed curtain wall of bronze mirror glass with horizontal mullions emphasized

FA-41 全玻璃幕墙，金属框架窗户
Fully glazed curtain wall, windows with metal frames

FA-42 光电嵌板覆层，现代商业大厦
Photovoltaic-panel cladding, modern commercial building

幕墙

摄影: Austin, Patterson, Disston

FA-43　普通砌层的新阿拉斯加黄杉板，墙角细部
New Alaskan yellow cedar shakes in common coursing, corner detail

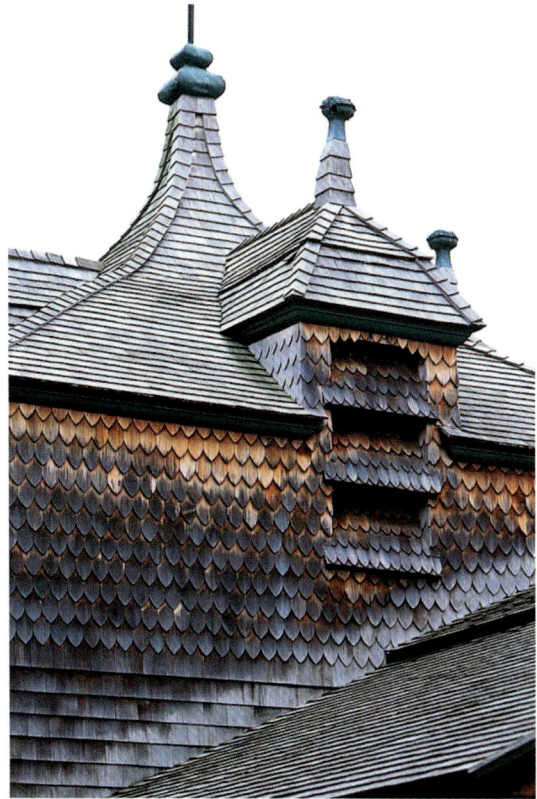

FA-44　鱼鳞图案双弯形屋顶上的风化犬齿贴面板
Weathered sawn shingles in fish-scale pattern on ogee roof

FA-45　带角板的鱼鳞图案墙板和锯齿线脚，平脚边和弦月窗
Clapboard and sawn siding in fish-scale pattern with corner board, flat applied trim, and lunette window

FA-46　交错层列的喷漆板
Painted shakes in staggered course

墙面板

FA-47 鱼鳞图案的喷漆贴面板，山墙细部
Painted sawn shingles in fish-scale pattern, gable detail

FA-48 屋顶窗上的鱼鳞喷漆贴面板
Painted sawn shingles in fish-scale pattern on shed dormer

摄影: Austin, Patterson, Disston

FA-49 普通层砌的风化阿拉斯加黄杉板和弧形拱窗，墙角和屋顶细部
Weathered Alaskan yellow cedar shakes in common coursing and segmental arch window, corner and roof detail

摄影: The Preservation Society of Newport County

FA-50 山墙上的木瓦；顶部：波浪图案环绕着牛眼窗；底部：鱼鳞图案
Cut shingles on gable. Top: wave pattern around bull's-eye window; bottom: fish-scale pattern

FA-51 左：饰有荷叶边屋顶板的普通层砌的木瓦；右：木瓦、鱼鳞和扇形交替图案；木瓦板
Left: shakes in common coursing with scalloped cut shingles; right: cut shingles, alternating fish-scale and scalloped patterns; shake border

摄影: Austin, Patterson, Disston

FA-52 普通层砌的新阿拉斯加黄杉木瓦，屋顶窗；复斜屋顶
New Alaskan yellow cedar shakes in common coursing, eyebrow window; gambrel roo

墙面板

FA-53　菱形图案的沥青墙面板（顶部），交错层列（底部）
Bituminous shingles in diamond pattern (top), staggered course (bottom)

FA-54　托架中楣柱下面的交错层列中的沥青墙面板
Bituminous shingles in staggered course below bracketed cornice

FA-55　人造琢石墙面图案中的沥青墙面板
Bituminous shingles in faux ashlar stone pattern

FA-56　人造砖图案中的风化沥青，裸露木质面板
Weathered bituminous shingles in faux brick pattern, exposed wood cladding

FA-57　多色六角形图案中的沥青墙面板
Bituminous shingles in multicolored hexagonal pattern

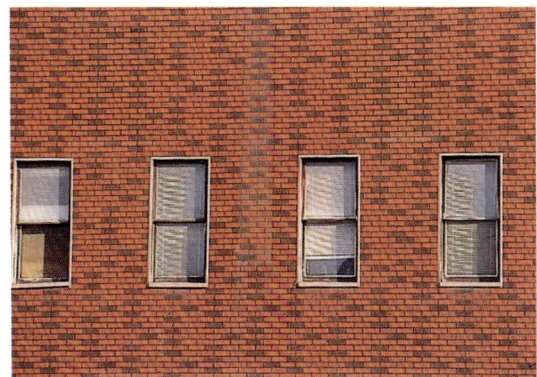

FA-58　仿砌砖式排列的沥青墙面板
Bituminous shingles in faux brick pattern

墙面板

FA-59 饰有彩色上釉装饰细节的砖石外立面：托架上楣、螺旋形装饰、花冠、肘状支柱

Brick facade with polychrome glazed ornamental details: bracket cornice, cartouches, garlands, ancon, and rope molding around semicircular windows

FA-60 有壁柱的华丽的石头饰面建筑面砖外立面

Ornate banded stone-finish architectural terracotta facade with pilasters

FA-61 带阿拉伯式图案装饰的彩色建筑面砖外立面

Polychrome architectural terracotta facade with arabesque ornament

摄影: Duane Langenwalter

FA-62 外墙上有龙图案的彩釉浮雕瓷砖

Glazed polychrome relief tiles with dragon design on exterior wall

FA-63 带有建筑面砖装饰的普通砖石外立面

Plain masonry facade with architectural terracotta ornament

FA-64 有彩色面砖装饰的砖石外立面

Brick facade with polychrome terracotta ornament

瓦和建筑陶瓷

FA-65　饰有阿拉伯式蔓藤花纹和漩涡装饰的镶嵌图案的檐部
Mosaic entablature with arabesque ornament and cartouche

FA-66　带有中东风格图案和整齐瓦窗的上釉瓷砖，庭院墙
Glazed tiles with Middle Eastern pattern and tile-trimmed windows, courtyard wall

FA-67　一个大理石壁柱旁边的大理石马赛克墙
Marble mosaic wall adjacent to a marble pilaster

瓦和建筑陶瓷

FA-68 嵌有玻璃砖窗口的麻面瓷砖，现代房屋入口
Textured matte tiles with glass-block windows, entry of modern house

FA-69 几何图案中配有水平瓷砖带的马赛克，现代公寓入口
Mosaic with horizontal bands of tiles in geometric pattern, entry of modern apartment house

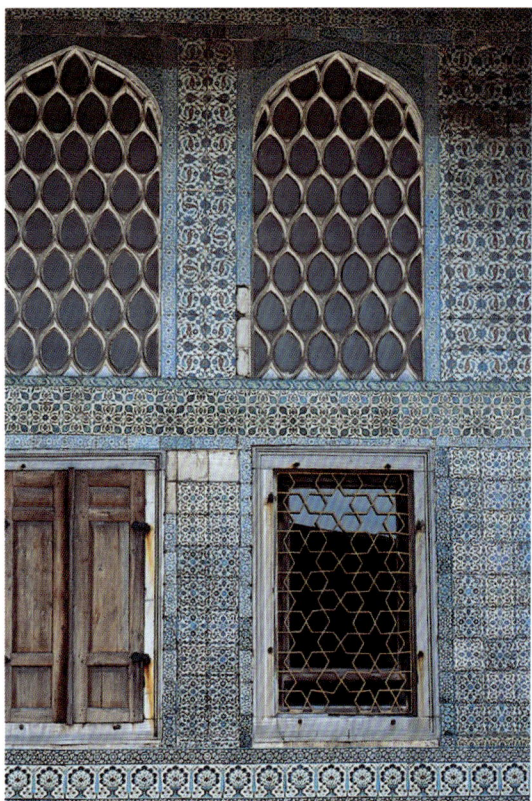

FA-70 带有图案的瓷砖围绕着有金属栅栏和百叶的方窗以及S形窗，土耳其
Glazed patterned tiles around square windows with metal grille and wood shutter and ogee windows with grille, Turkey

FA-71 墙面的几何图形彩色瓷砖和现代公寓的阳台
Polychrome tiles in geometric patterns on walls and balconies of modern apartment house

瓦和建筑陶瓷

摄影: Meredith Barchat

FA-72 拉毛水泥墙、门和活动式窗板的传统亮色喷漆，中美洲
Traditional bright colors painted on stucco walls, door, and shutters, Central America

FA-73 传统红色喷漆木料填充的梁柱外立面和门板，日式入口的大门
Traditional red painted wood-infill post-and-beam facade and paneled door, Japanese entry gate

FA-74 商店外部有彩喷几何设计和金属雕刻元素的砖墙
Brick wall with painted geometric design and painted metal sculptural elements on store exterior

彩绘墙

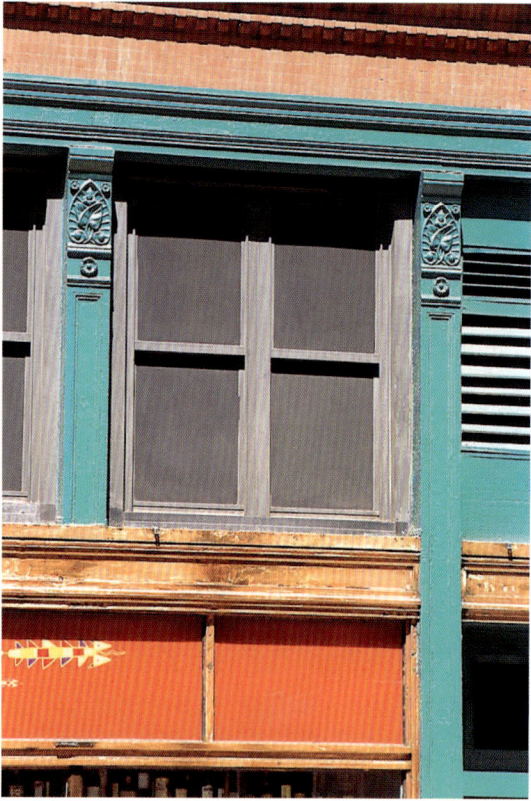

FA-75 非传统彩喷的有横楣的铸铁外立面
Trabeated cast-iron facade painted in nontraditional colors

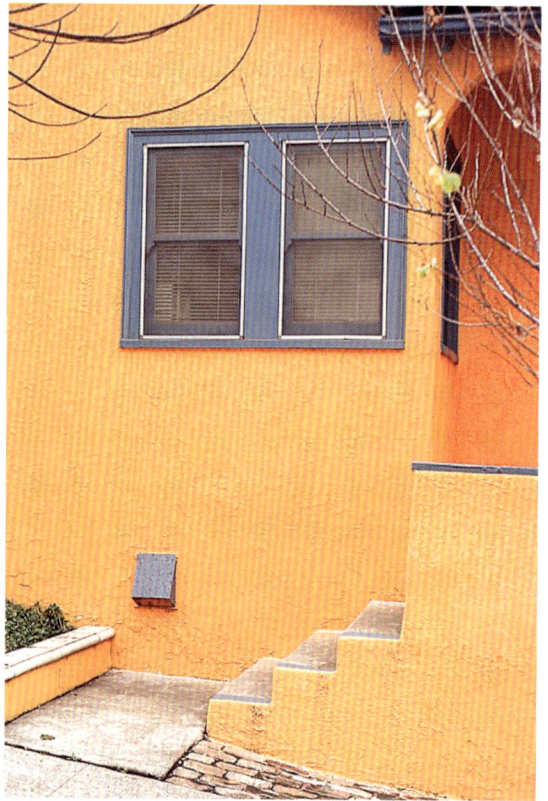

FA-76 窗四周有对比色的喷漆织纹拉毛水泥，现代房屋
Painted textured stucco with contrasting color around window,
contemporary house

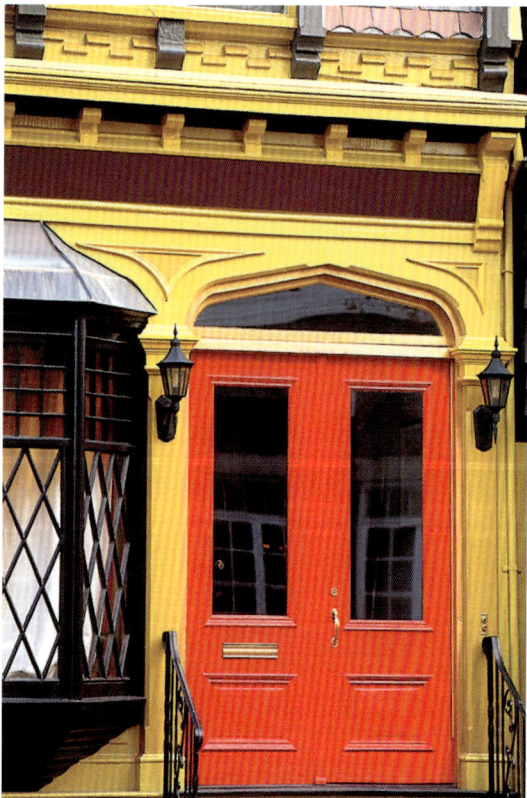

FA-77 都铎式拱门和相临的飘窗被喷涂为鲜艳的颜色，19世
纪的商店店面
Tudor-arch doorway and adjacent oriel painted with exterior gloss
colors, 19th-c. storefront

FA-78 有窗和过梁的外立面上喷有花纹装饰带的拉毛水泥砖
坯，18世纪美国西南部地区
Stucco-covered adobe painted with floral and ornamental bands
on facade with window and exposed lintel, 18th-c. American
Southwestern mission

彩绘墙

FA-79　有八角塔的彩色房屋，陡峭的山形墙顶窗，突出的盖板、鱼鳞式侧面护墙板和飞檐托饰，19世纪
Polychrome house with octagonal turret, steep gable dormer, prominent bargeboards, fish-scale cut siding, and modillion cornice, 19th-c.

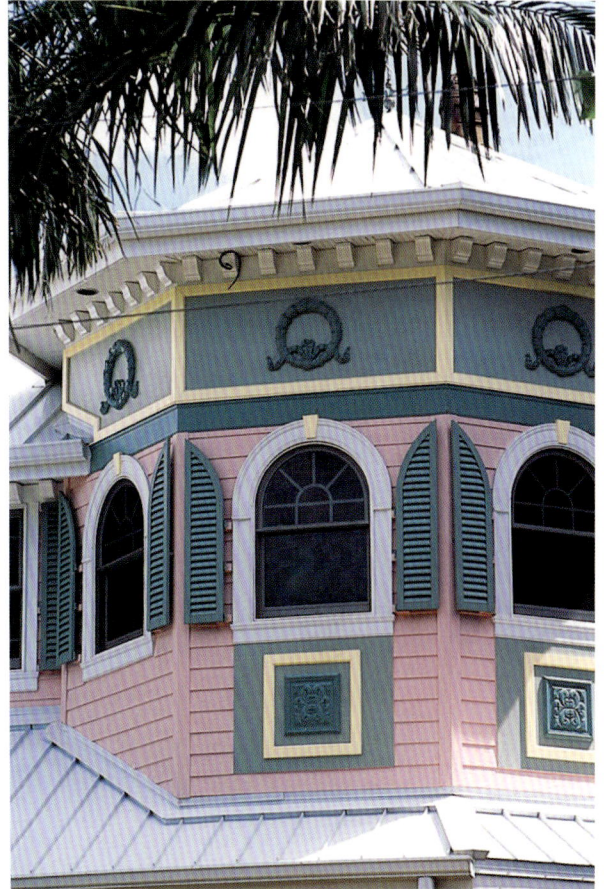

FA-80　当代彩色房屋的八角塔，带有飞檐托饰、重复装饰和窗户四周的装饰性百叶窗
Octagonal turret on contemporary polychrome house, modillion cornice, repeated ornament, and ornamental shutters around windows

FA-81　山形墙尾端有小块挡板的彩喷车库，企口墙板和门
Painted garage with shingles on gable end, vertical tongue-and-groove siding and door

彩绘墙

FA-82　风化的绘制角石和托斯卡纳墙上的石头细部
Weathered trompe l'oeil quoins and stone detail on Tuscan wall

FA-83　托斯卡纳百叶窗四周的风化的绘制线脚和三角形楣饰
Weathered trompe l'oeil moldings and pediment around shuttered windows on Tuscan wall

FA-84　彩喷的石质庭院墙和带有雕刻线的人造石结构的混凝
　　　　土砖，印度尼西亚
Courtyard wall of painted stone and concrete block with scribed lines simulating masonry, Indonesia

FA-85　庭院墙和台阶：有彩喷凹槽和装饰的风化壁柱
Courtyard wall and steps: weathered pilasters with painted flutes and ornament, Indonesia

FA-86　突拱四周的叶形装饰喷绘和托斯卡纳拉毛水泥墙上的
　　　　窗户
Decorative painting of foliated ornament around rusticated arch and window on stucco Tuscan wall

FA-87　墙面上的五彩拉毛粉饰壁柱、丰富的拱门饰、古典线
　　　　脚和网状花格装饰，意大利
Sgraffito depicting pilasters, enriched archivolts, classical moldings, reticulated band with paterae on wall, Italy

彩绘墙

FA-88 工业建筑砖砌外立面上的彩绘几何图案（涂料制造厂）
Painted geometric pattern on brick facade of industrial building (paint manufacturer)

FA-89 整修中的建筑的彩喷胶合板覆盖层
Multicolored painted plywood cladding on building under renovation

FA-90 带犬齿隔带的砖砌外立面，上面绘制有云和天空的图案
Trompe l'oeil clouds and sky on brick facade with dog-tooth courses

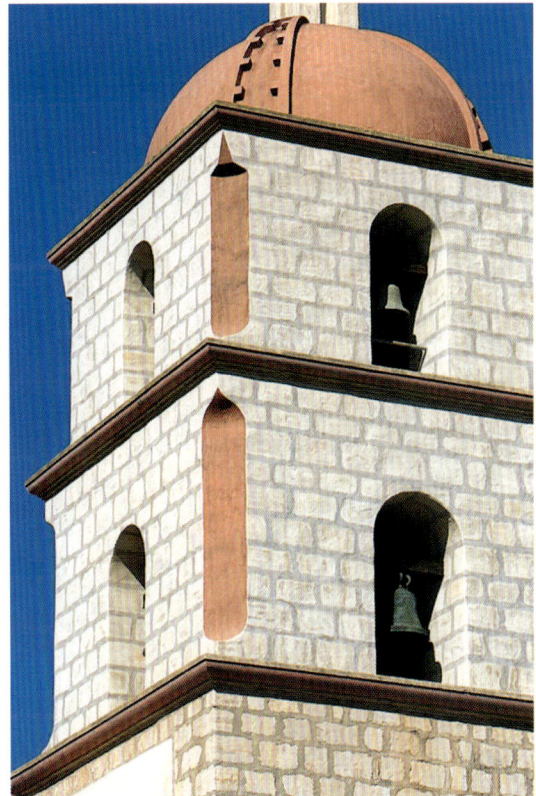

FA-91 钟楼上的绘制人造砖石，有削角的圆顶，19世纪建筑
Trompe l'oeil stucco simulating adjacent masonry on belfry, dome with chamfered corners, 19th-c. mission

彩绘墙

FA-92 裸露的砖墙上涂有已风化的涂色拉毛水泥，上部是半
木结构的山形墙
Weathered painted stucco over exposed brick with half-timber
gable above

FA-93 彩喷拉毛水泥外立面，意大利临河建筑
Painted stucco facades, Italian riverfront buildings

FA-94 风化喷漆护墙板外立面
Weathered painted clapboard facade

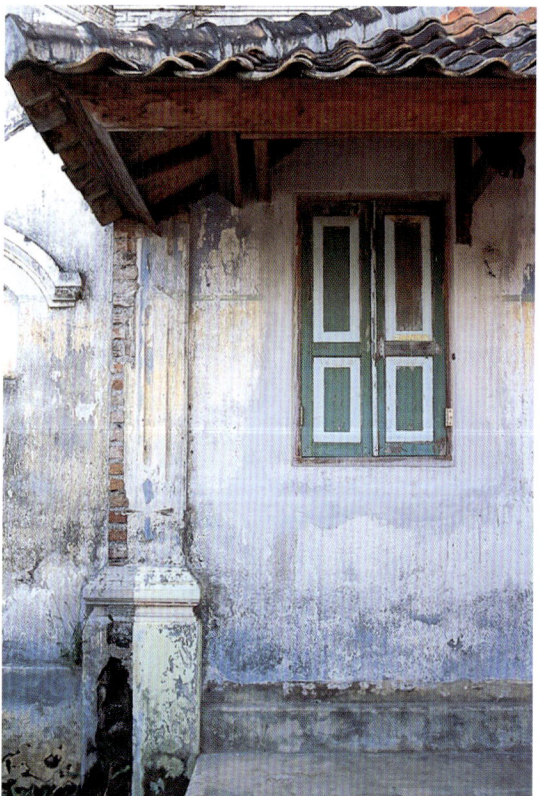

FA-95 砖石表面的风化喷涂拉毛水泥
Weathered painted stucco over brick

彩绘墙

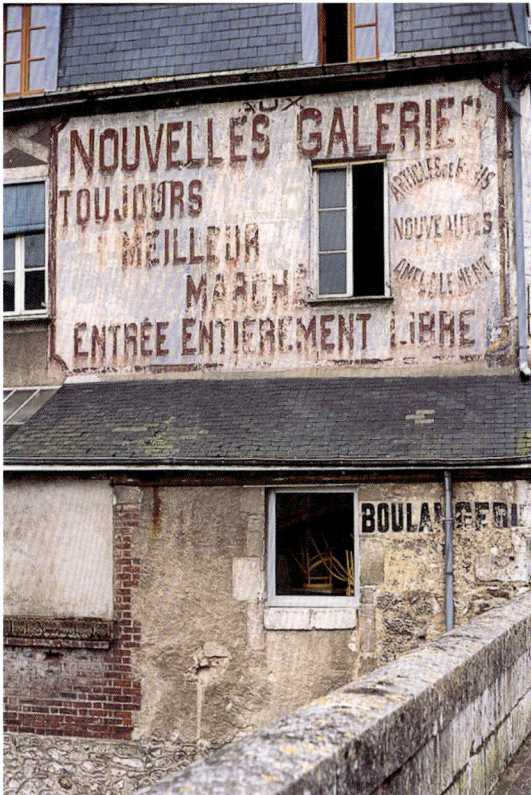

FA-96 拉毛水泥墙上的喷涂广告，法国
Painted advertisement on stucco wall, France

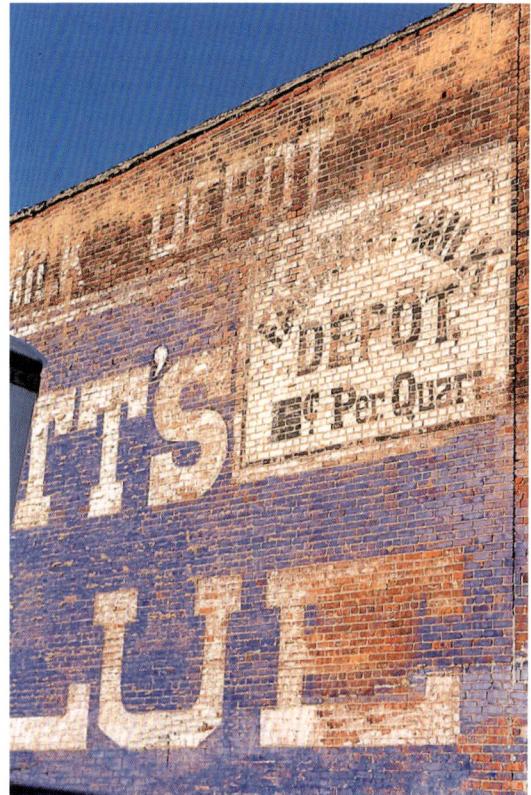

FA-97 风化的都市砖墙上被风雨侵蚀过的喷涂广告
Weathered painted advertisement on weathered urban brick wall

FA-98 带有刮痕的仿砖石接缝的现浇筑混凝土框架上的剥漆涂料
Peeling paint on poured-in-place concrete frame with scored simulated masonry joints

FA-99 有石块基座的砖质花园墙上的剥漆涂料
Peeling paint on brick garden wall with stone base

FA-100 风化喷涂垂直板和板条墙以及相邻的水泥砖石结构
Weathered painted vertical board-and-batten wall and adjacent masonry structure

FA-101 风化喷漆护墙板和对角板框架门的线脚
Weathered painted clapboard and siding with diagonal board framed door

彩绘墙

FA-102　砖墙上涂鸦风格的标志性喷涂和波浪钢城市墙
Sign painted in graffiti style on brick and corrugated metal urban wall

FA-103　砖砌外立面上的喷涂广告
Painted advertisement on brick facade

摄影：Duane Langenwalter

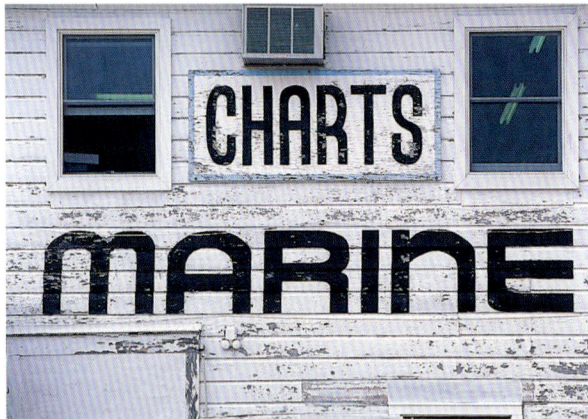

FA-104　喷漆舌槽式外立面上的喷漆标志
Painted sign on painted tongue-and-groove facade

FA-105　有升降门的砖砌外立面的喷涂标志
Painted sign on brick facade with loading door

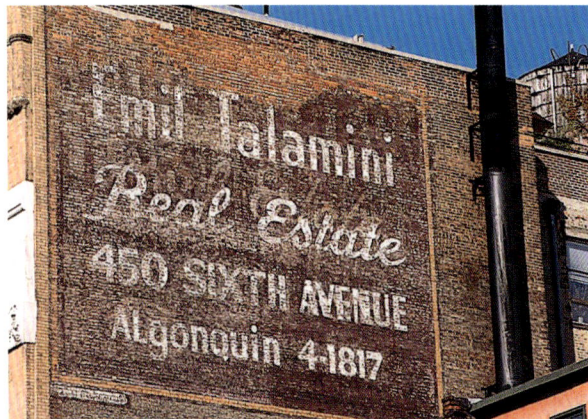

FA-106　空白砖砌外立面的喷涂标志
Painted sign on blank brick facade

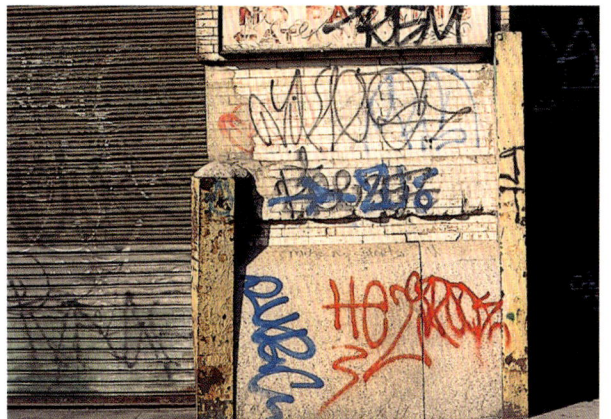

FA-107　车库门之间砖墩上的涂鸦
Graffiti painted on brick pier between garage doors

彩绘墙

FA-108　哥特风格的彩色混合拱入口细部，配有三个侧柱
Detail of polychrome compound-arch portal in Gothic style, with three jamb shafts

FA-109　有啮合柱和螺旋形栏杆的梯塔
Stair tower with engaged columns and spiral balustrade

FA-110　带有传统石雕和奇异形状塑像的砖砌外立面，印度尼西亚寺庙
Brick facade with traditional stone carving and grotesque figures, Indonesian temple

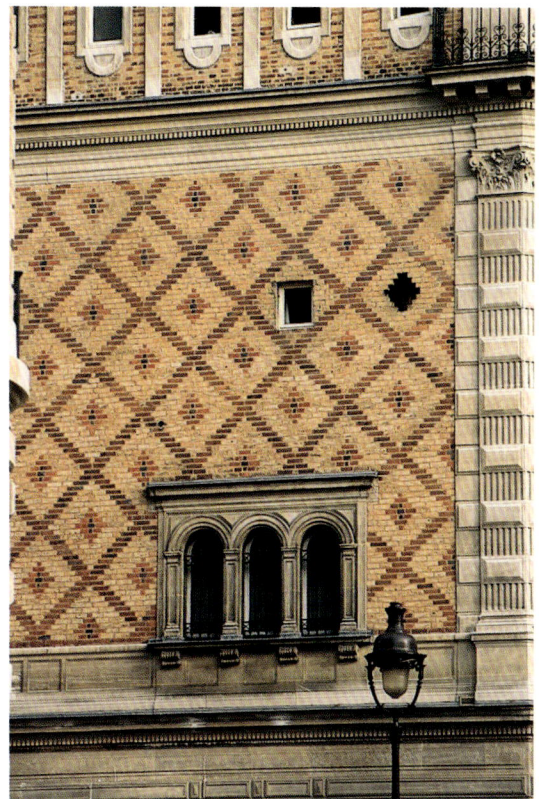

FA-111　菱形图案的砖砌外立面，以凸角壁柱和檐口为框架
Diamond-patterned brick facade with window arcade, framed by a rusticated corner pilaster and cornice

立 面 装 饰

外立面

FA-112 模制上檐口和多层砖石外立面上被波状装饰围绕的窗户
Molded cornices and window surrounds with running ornament on
multistory masonry facade

FA-113 正门上的阳台，分开的楣饰和多层砖石外立面上华丽的装饰
细部
Balconies over portal, broken pediment, and ornate details on multistory
masonry facade

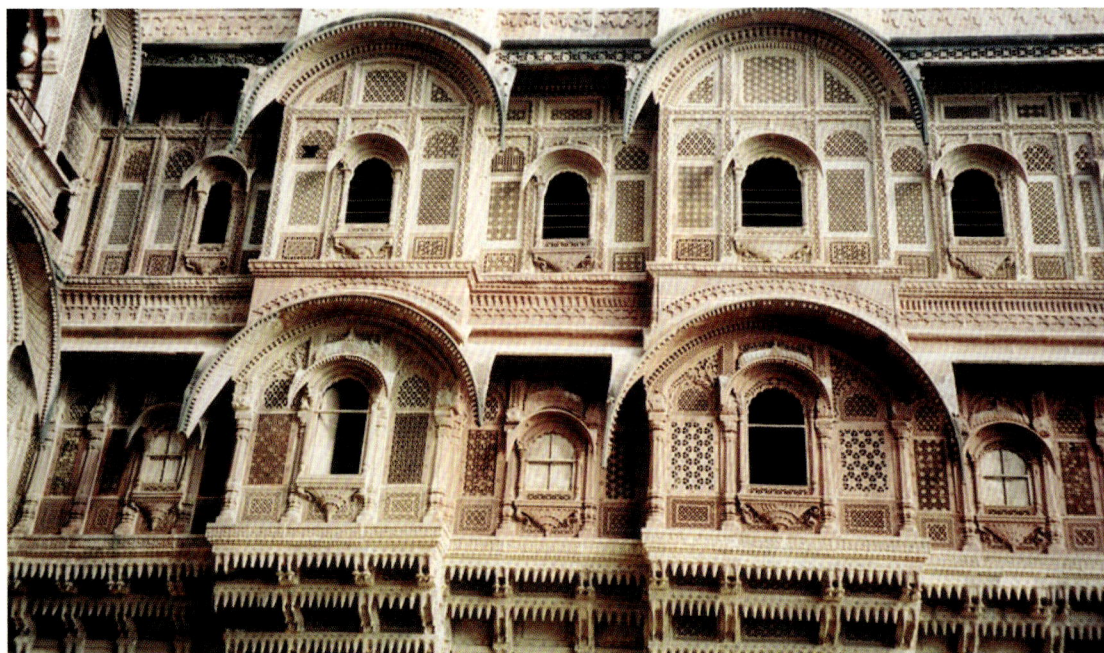

FA-114 多层外立面上镂空木雕的华丽阳台，印度
Ornate balconies of carved and pierced wood on multistory facade, India

摄影：Meredith Barchat

立面装饰

FA-115　拉毛水泥外立面，较古老的拱窗嵌入墙中，旁边是现代窗户
Stucco facade with older arches embedded in the wall adjacent to modern windows

FA-116　装饰艺术风格外立面，上釉砖下方有几何图案的精细打磨的砖面板
Art Deco facade, highly polished stone cladding with geometric ornament below white glazed brick

FA-117　砖石外立面中有装饰拱腹的新哥特式装饰元素
Neo-Gothic ornamental features with ornamental spandrels in brick facade

FA-118　有石块或浇筑公羊头和蛇装饰的砖砌外立面，以及上下层窗间墙面上的波状几何装饰
Brick facade with stone or cast rams-head-and-snake ornaments and running geometric ornament in spandrel bands

立面装饰

FA-119 窗口间带有装饰性带状雕刻拱肩装饰的砖砌外立面
Brick facade with decorative masonry frieze of sculpted spandrels between windows

FA-120 拱肩上带有阿拉伯式浮雕叶形装饰的石砌外立面
Stone facade with relief foliated arabesque ornament on spandrels

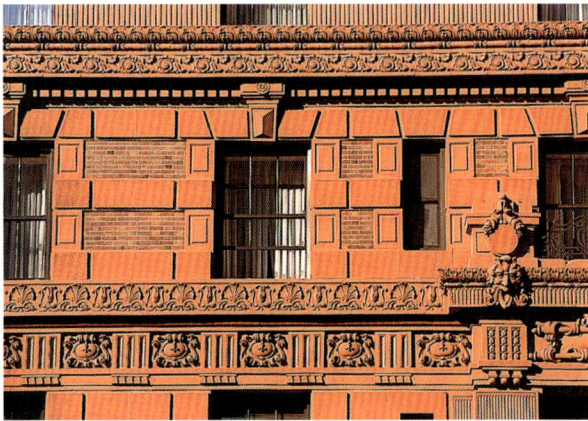

FA-121 由带有粗面繁重装饰的褐色砂石或者面砖组成的砖砌外立面
Brick facade with rusticated and heavy embellishment in brownstone or terracotta

FA-122 带有繁重叶形浮雕图案嵌板的外立面
Facade clad in heavily embellished relief panels

FA-123 有双啮合柱的正门和用椭圆形图案装饰和美化的阳台，18世界英国乡村房屋
Portal with coupled engaged columns and balcony heavily ornamented and embellished with cartouches, 18th-c. English country house

立面装饰

FA-124 带有拉毛水泥和浇筑混凝土浅浮雕的彩色外立面，
装饰艺术风格旅馆
Polychrome facade with low relief ornament in stucco and cast
concrete, Art Deco hotel

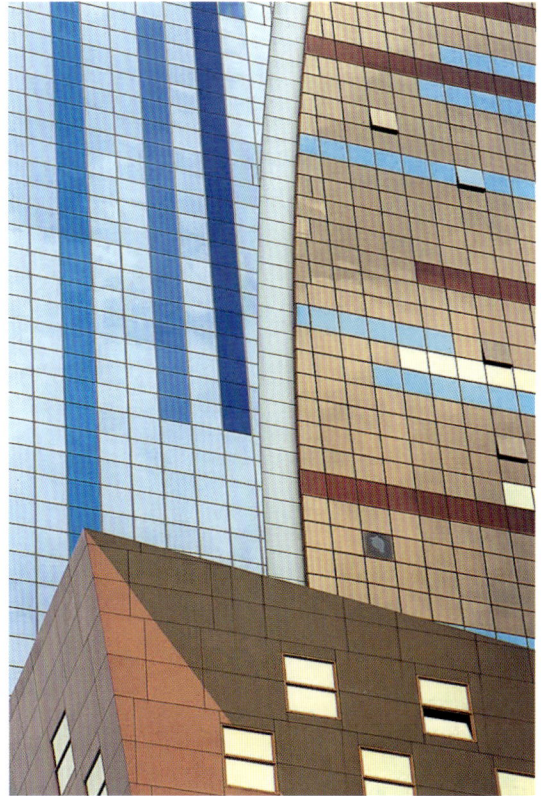

FA-125 彩色嵌板的幕墙系统，带有大型图案，现代商业大
厦
Curtain-wall system of multicolored panels, supergraphic,
modern commercial building

FA-126 弯曲玻璃幕墙，由简明石墩和锁孔窗环绕的石头贴
面外立面，现代商业大厦
Curving glass curtain wall above stone-veneer facade with
expressed piers and keyhole window surrounds, modern
commercial building

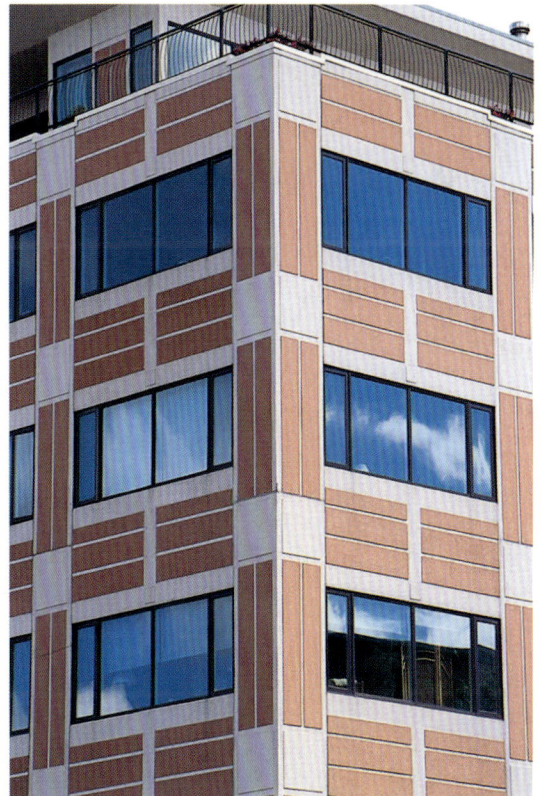

FA-127 彩色交叉结构的带横梁的外立面，现代多功能建筑
Trabeated facade finished in colors suggesting post-and-lintel
construction, modern mixed-use building

立面装饰

外 立 面

FA-128 饰有明亮和半透明玻璃嵌板的幕墙和有突出边缘及
竖框的柱形盖
Curtain wall with geometric design in clear and opaque glass
spandrels and column covers with raised edges and mullions

FA-129 带有多种镶边面板的幕墙，装饰有通风格子窗和突
出的金字塔形以及方形套圆形图形
Curtain-wall panels in multiple bands with ventilation grilles
incorporated in projecting ornamental pyramid and circle-in-
square motifs

FA-130 带有简式冲压窗、弯曲的阳台和大比例栅格的外立
面，现代公寓
Facade of simple punched windows, curvilinear balconies, and
large-scale open grid, modern apartment house

FA-131 带有彩色填充和上下窗间墙面嵌板的有横梁的外立
面，现代商业大厦
Trabeated facade with multicolored infill and spandrel panels,
modern commercial building

立 面 装 饰

装 饰 和 线 脚

OM-1 镶边，从顶部到底部：三角形和玫瑰花形雕饰；有星星和玫瑰花形雕饰的方形内置嵌板；包金的叶形装饰图案、垂直小凸嵌线和图案
Bands. Top to bottom: triangles and rosettes; square inset panels with star and rosette; gilded foliated motif, vertical reeding, dart motif

OM-2 中心有贝壳图案环绕的玫瑰花形铜铸八角保险箱
Cast bronze octagonal coffer with central rosette surrounded by shell-motif molding

OM-3 基于上横梁柱顶盘帕戈萨斯塑像（"墙面"）同"火炬"装饰（"竖条纹饰"）交替排列
Cast bronze frieze based on classical entablature: Pegasus figures ("metopes") alternating with "torch" ornaments ("triglyphs")

建筑装饰

OM-4 镶边：从顶部到底部：配有贝壳檐口装饰的格式化的维特鲁威风格的卷形石雕饰物；垂直的波浪形图案；绳索状的线脚，和由中心向左右两边变化的叶形装饰线脚
Bands. Top to bottom: stylized Vitruvian scroll with shell antefix; vertical corrugated pattern; rope molding, and foliated ovolo molding with pattern reversed at center

OM-5 卷轴形铜铜质山形墙上的螺旋形雕饰；从顶部到底部的线脚包括：有镶边的挑檐滴水板；卵锚馒形装饰；舌镖图案装饰板条
Volute on bronze scrolled pediment. Molding, top to bottom: corona with band ornament; egg-and-dart ovolo; tongue-and-dart cyma reversa

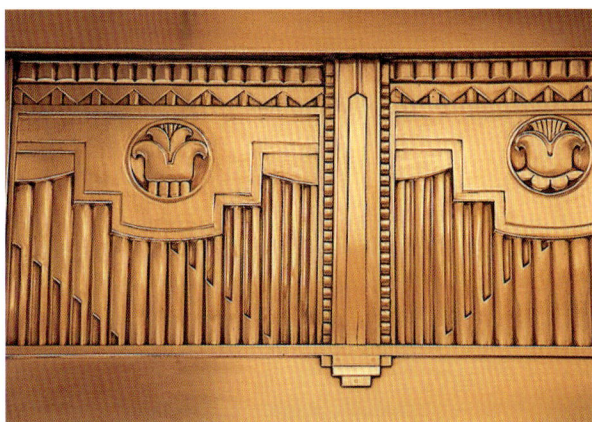

OM-6 饰有垂直小凸嵌线的铜面板，其几何轮廓下有由花形装饰和带状几何装饰组成的装饰框
Bronze panels with vertical reeding below geometric profile with floral medallions and banded geometric ornamented frame

OM-7 有花状平纹的叶尖饰涡卷装饰的铜质山形墙细部
Detail of bronze scrolled pediment with anthemion finials

OM-8 海马雕饰的镀镍银层铜或浇铸不锈钢嵌板
Nickel-silver-plated bronze or cast stainless steel spandrel panels with seahorse ornament

建筑装饰

OM-9 有怪异装饰的雕木悬壁拖梁支柱
Carved wood ancon with mascaron

OM-10 龙形的奇特造型托架
Antic bracket in the form of a dragon

OM-11 托架：卷形雕饰前有狮子头和花环围绕的方形奇特嵌板
Brackets: square panel surrounded by a garland with a lion's-head
antic in front of a scroll

OM-12 带有茛苕叶形装饰的螺形支架
Console with acanthus leaves

建筑装饰

OM-13 入口上方的叶形装饰，侧翼带有之字形（顶部）和锯齿形镶边（底部）
Foliated ornament over doorway flanked by zig-zag (top) and dentilated (bottom) bands

OM-14 由花和织物装饰的无筋砾石墙角细部
Corner detail of plain tablets with festoons and drapery ornament

OM-15 饰有茛苕叶形装饰的螺形支架，侧翼饰有花彩
Console with acanthus leaves, flanked by festoons

建筑装饰

OM-16 奇异的装饰，印度尼西亚
Antic ornament, Indonesia

OM-17 雕刻的都铎式蔓叶花样饰在无筋壁柱拐角
Sculptural vignette set in a plain plastered corner

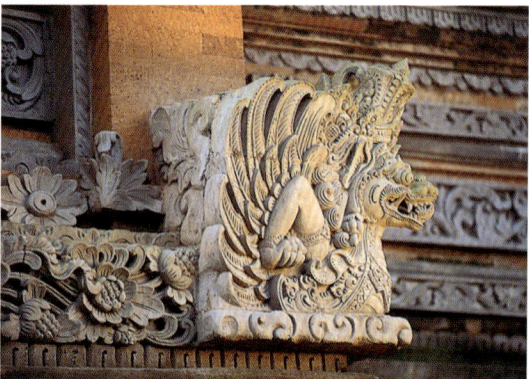

OM-18 怪诞的墙角装饰，印度尼西亚
Antic corner ornament, Indonesia

OM-19 装饰，半狮半鹫的怪兽手持一本书
Ornament, griffin holding a book

OM-20 彩色浮雕，树上的猴子，日本
Polychrome relief carving, monkeys in a tree, Japan

摄影: Duane Langenwalter

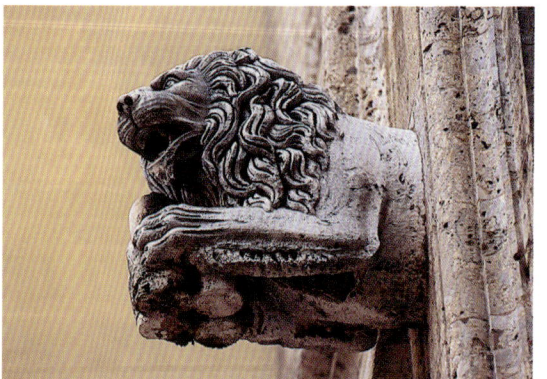

OM-21 狮子滴水嘴
Lion gargoyle

建筑装饰

OM-22 刻有建筑年份的椭圆形轮廓装饰
Cartouche with building date

OM-23 装饰有武器的漩涡形装饰，周围环绕着旗帜、大炮、鼓和横笛
Cartouche with regimental arms surrounded by flags, cannons, drums, and fife

OM-24 上面带有楼牌号的螺旋形图案装饰，饰有花和彩带，中心有奇特的图案
Cartouche with building number above guttae, with festoon and center antic

OM-25 有叶形装饰的门楣，带有镀金楼牌号的凹面椭圆形装饰
Tympanum with foliated ornament, concave oval cartouche with gilded building number

OM-26 螺形支架上刻有建筑竣工的年份和鸟形浮雕装饰
Consoles flanking a date stone with bird ornament in relief

OM-27 饰有花和彩带的无筋楣石，上面有刻有楼牌号的凸面椭圆形装饰
Convex oval cartouche with building number over a plain lintel with a festoon

建筑装饰

OM-28 有拉丁文碑文的风化石碑
Weathered stone tablet with Latin inscription

OM-29 方形嵌板里面有一个旭日形装饰的砂石椭圆轮廓装饰
Sandstone cartouche in square panel with a mascaron inside a sunburst

OM-30 有阿拉伯文碑文的涡卷饰，被叶形装饰所环绕
Cartouche with Arabic inscription, surrounded by foliated ornament

OM-31 上方顶着王冠的大涡卷饰，带有窗耳的线脚框，被相似设计的小漩涡饰所环绕
Large cartouche surmounted by a crown, with molded frame with crossettes, surrounded by smaller cartouches of similar design

OM-32 入口上方有三个涡卷饰，中间的一个被两个天使包围
Three cartouches over an entryway, the center one flanked by two cherubs

建 筑 装 饰

OM-33 有蛇形图案和八边形框中的玫瑰花形雕饰
Ornamental band with serpentine motif and rosette in an octagonal frame

OM-34 希腊回纹式图案中饰有玫瑰图案的装饰，位于卵锚饰线脚上方
Ornamental band with rosettes in a Greek-key motif, over egg-and-dart molding

OM-35 封闭栏杆下带有维特鲁威风格卷轴形石雕的装饰带
Ornamental band with Vitruvian scroll below blind balustrade

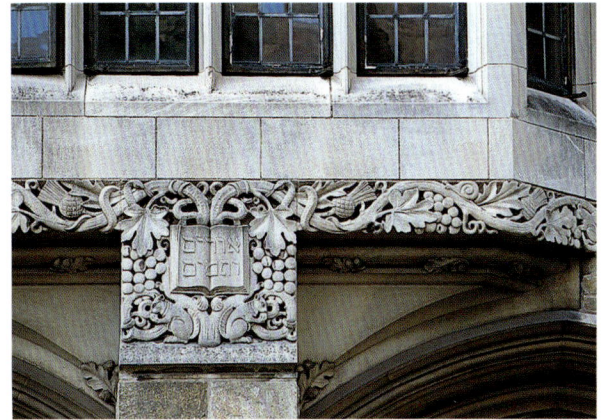

OM-36 飘窗下的托臂底边上带有叶形装饰和刻有希伯来文的翻开的书的形状所装饰
Foliated ornament and open book with Hebrew inscription on base course of a corbel under an oriel

OM-37 彩色装饰带：有圆形图案的宽中间带，两侧有直线窄边图案；底部附加叶形装饰带
Polychrome ornamental bands: wide central band with a circle motif, flanked by thin bands with linear patterns; foliated band on the bottom

OM-38 花园墙遮檐下连接着棕榈树图案的装饰带
Ornamental band with interlocking palm motif below the coping of a garden wall

建筑装饰

摄影: The Preservation Society of Newport County

OM-39 带有怪状头像的墙壁托架，左边有天使和叶形雕刻嵌板
Shaft on a mascaron bracket. At left, foliated sculptural panel with a cherub

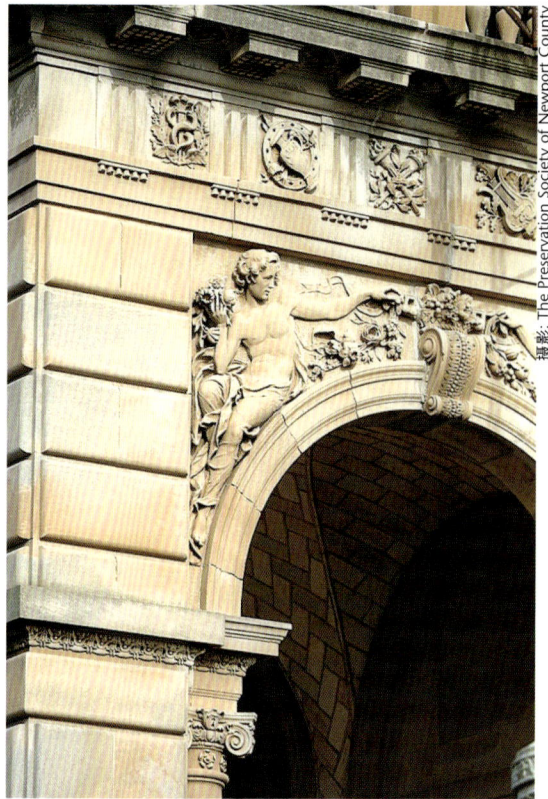

OM-40 有雕刻图案拱肩的入口，陶立克式柱顶部带有排档间饰三竖线花纹装饰，下面是带有拱心石的悬臂托梁
Portal with sculpted spandrel, ancon keystone below Doric-style entablature with sculpted metopes and triglyphs

OM-41 墙面扶壁上的造型尖顶饰
Figurative finial on a wall buttress

OM-42 壁端柱饰有手持棕榈树叶长翅膀的神话人物的浮雕，檐口下方饰有卵锚饰圆凸线脚和犬齿与飞檐托饰相交替出现的镶边
Anta with relief winged mythological figure holding palm fronds below cornice with egg-and-dart ovolo molding and an alternating dentil and modillion band

OM-43 哥特式风格入口处由祭坛华盖覆盖着的壁龛
Niche covered by a baldachin on Gothic-style entry

OM-44 寺庙入口有奇特人物雕像的梯状叶形装饰图案，巴厘岛
Stepped and foliated ornamental pattern with antic figure on temple entry, Bali

OM-45 被奇特人像包围的椭圆形轮廓装饰，各种各样的线脚，从顶部到底部分别是：叶形装饰的反向波状花边；叶形装饰的正向波状花边；卵锚形圆凸形线脚装饰；舌锚形线脚；凸圆线脚
Cartouche surmounted by antic. Enriched molding, top to bottom: foliated cyma reversa; foliated cyma recta; egg-and-dart ovolo; tongue-and-dart; bead-and-reel

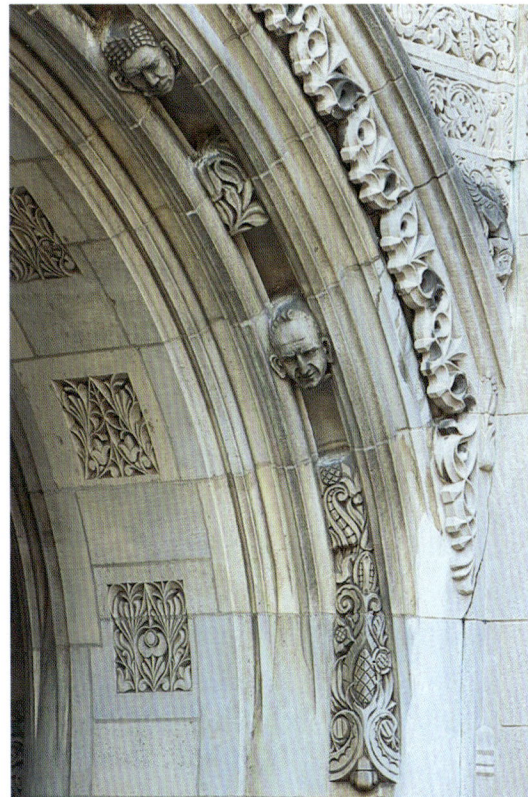

OM-46 拱门饰上有装饰有怪异头像和植物装饰的凹式线脚
Mascarons and botanical ornaments on a concave molding on an archivolt

建筑装饰

OM-47 有中美洲风格图案的叶形装饰柱身
Foliated shaft with Mesoamerican-style mascaron

OM-48 风化石门四周带有莨苕叶形图案的装饰
Foliated ornament with acanthus leaves on weathered stone door surround

OM-49 镀金装饰：大奖章、叶形装饰带、带奇怪头像的托架和中心叶尖饰
Gilded ornament: large medallions, foliated band, bracket with a mascaron, and central finial

OM-50 雕刻有怪异头像的拱心石
Carved stone mascaron keystone

OM-51 入口上方带有浮雕的中楣：周围有人物雕像装饰的牛眼窗，卷轴形图案和扁平壁柱上的混合式柱冠
Stone relief frieze over entry: bull's-eye window framed by draped figures, scroll motifs, and Composite capitals on flat pilasters

OM-52 带有雕刻拱肩的拱门，螺形支架间的柱顶线盘饰有垂花雕刻和带状装饰，尽头是位于线脚下的圆锥装饰
Arch with sculpted spandrel. Entablature with frieze and festoons between consoles terminating in guttae below molding

建筑装饰

OM-53 简式檐口上方的不完整卷轴山形墙和刻有文字的横楣
Stylized broken scrolled pediment over a plain cornice and inscribed architrave

OM-54 被玫瑰花图案环绕的门；墙角带有花状平纹图案和花线装饰的凹雕檐口；由花状平纹装饰的方柱
Door surround with rosettes; cavetto cornice with anthemion motif and antefix at corner; pier capped by anthemion

OM-55 被希腊回纹式图案环绕的窗耳的悬臂托梁，另一端是三个圆锥装饰；拱肩嵌板中有菱形设计外框
Crossette surround with Greek-key motif, terminated with three guttae; framed lozenge design in spandrel panel

OM-56 左边：卷轴式拱心饰一端为叶形装饰拱墩；右边：纽索状装饰图案的嵌条
Left: scrolled archivolt terminating at a foliated impost; right: banding with guilloche pattern

建筑装饰

OM-57 有方形花图形和希腊回纹饰图形的镶边；侧柱的线脚带由凸圆线脚包围的玫瑰图案
Greek-key band with square floral element; jamb molding with rosettes framed by ornamented ovolo molding

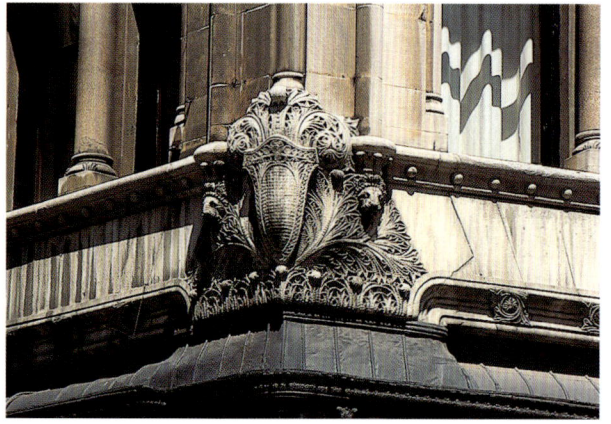

OM-58 叶形装饰的墙角细部，带有遁形图案和奇形怪状的图案
Corner detail of foliated ornament, shield device, and grotesque

OM-59 入口处檐口的花线装饰，中间是一个怪异头像
Antefixes on cornice over a portal, the center one a mascaron

OM-60 连拱廊中心饰有狮子头像的叶形装饰阿拉伯式中楣
Foliated arabesque frieze with central lion's head over an arcade

OM-61 从顶部到底部：壁龛上的椭圆轮廓装饰；不同的带有玫瑰图案的四瓣花图案交替出现的装饰嵌板；哥特式线脚
Top to bottom: cartouche over a niche; alternating ornamental panels with quatrefoil devices with rosettes; Gothic molding

OM-62 檐口上反向重复的顺砖装饰，从中间向两侧分别是：花状平纹、涡卷饰、火炬、涡卷饰、叶形装饰、简朴的皿形饰
Reverse and repeat running ornament above a cornice: from center out: anthemion, volute, torche, volute, foliated ornament, plain patera

OM-63 右下：维特鲁威风格的卷形石雕；左上：卵锚形线
脚；装饰线脚上下均有饰带
Bottom right: Vitruvian scroll; upper left: egg-and-dart molding.
Fascia above and below ornamented molding

OM-64 石头线脚，从顶部到底部分别是：卵锚饰、齿状饰带、
普通饰带、卵锚凸圆形线脚、叶形装饰带、树叶—箭形
或心形—箭形装饰、半圆式线脚间的普通饰带
Stone molding. Top to bottom: egg-and-dart, dentil band, fascia,
egg-and-dart ovolo, foliated band, leaf- or heart-and-dart, fascia
between two bead-and-reel moldings

OM-65 螺形支架上的檐口，从顶部到底部分别是：花装平纹
正向线脚、树叶—箭形反向波纹线脚、飞檐托饰饰
带、卵锚形凸圆线脚和齿状饰带
Cornice on console. Top to bottom: anthemion cyma recta, leaf-
and-dart cyma reversa, modillion band, egg-and-dart ovolo
molding, dentil band

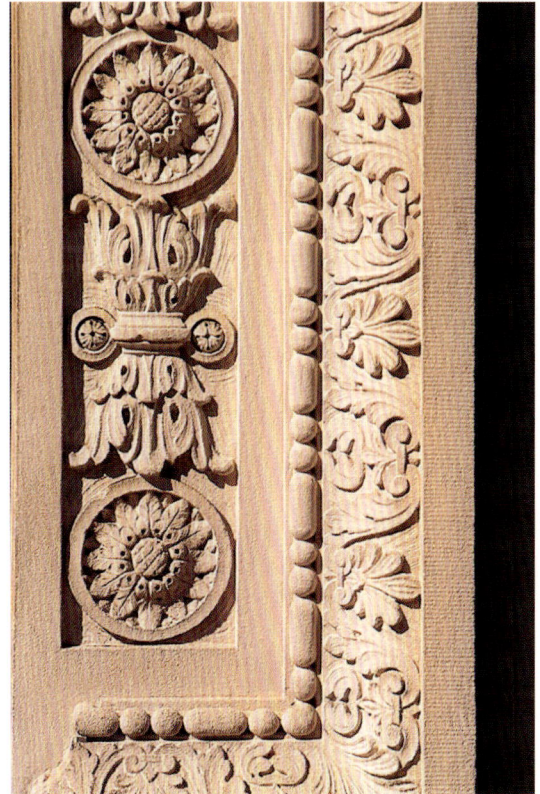

OM-66 建筑一角的细部，带有蜿蜒的叶形装饰的凸圆线脚；
叶形设计外围是半圆饰，侧面饰有玫瑰图案
Corner detail of enriched ovolo molding with trailing foliated pattern; bead-and-reel framing foliated design flanked by rosettes

线脚

OM-67 柱顶线盘，从顶部到底部分别是：树叶—箭形装饰、简式挑檐滴水板、反曲线装饰板条、齿状饰带、卵锚线脚、叶形装饰中楣、装饰性舌榫反曲线脚；柱冠部分：卵锚圆凸角线、半圆饰、花状平纹凹弧线脚
Entablature. Top to bottom: leaf-and-dart, plain corona, cyma, dentil band, egg-and-dart molding, foliated frieze, modified tongue-and-dart cymatium. Capital: egg-and-dart echinus, bead-and-reel, anthemion cavetto

OM-68 简洁的古典式基座，有座盘饰、突出横饰线、柱基的凹形边饰、座盘饰和方形基柱
Simple Attic base with torus, fillet, scotia, torus, plinth

OM-69 梯状饰带线脚、内置嵌板、托臂块体和类似托饰的元素
Stepped fascia moldings, inset panels, corbelled block, and modillion-like elements

OM-70 檐口从顶部到底部分别是：简式圆凸形线脚、齿状饰带、简式反曲线装饰、简式正波纹线装饰
Cornice. Top to bottom: plain ovolo molding, dentil band, plain cyma reversa, plain cyma recta

OM-71 科林斯式顶柱线盘，从顶部到底部分别是：有花状平纹的正波纹线、玫瑰图案和突出的装饰托架、卵锚形圆凸线脚、齿状饰带、叶形装饰带、舌锚反曲线形饰带
Corinthian entablature. Top to bottom: cyma recta with anthemion, rosette-and-modillion cornice, egg-and-dart ovolo molding, dentil band, foliated frieze, tongue-and-dart cyma reversa

OM-72 卵锚形圆凸饰线脚的墙角细部
Corner detail of egg-and-dart ovolo molding

OM-73 金属嵌板组成的中楣和檐口，墙角分开露出"檐口"拱腹的
"古典柱式顶部"，现代建筑外立面
Metal cladding suggesting a frieze and cornice, "entablature" broken at corner to reveal the soffit of the "cornice," modern building facade

OM-74 金属饰带组成的线脚，现代建筑外立面
Metal fascia bands suggesting molding, modern building facade

OM-75 金属饰带组成的古典的柱顶线盘，竖框的反复出现构成了飞
檐托饰带
Corner detail of metal fascia bands suggesting a classical entablature, the continuation of mullions suggesting a modillion band

OM-76 带有饰带的金属嵌板构成了凸圆线脚，现代外立面
Metal panel with fascia framing bead molding, modern facade

OM-77 铸造金属线脚，从顶部到底部分别是：正波纹线装饰、凹圆
线脚、梯状平线脚、凹弧饰、正波纹线装饰、反曲线装饰
Cast metal molding. Top to bottom: cyma recta, cove, stepped fillets, cavetto, cyma recta, cyma reversa

OM-78 金属饰带组成的古典柱式顶部，在间隔处断开的分界
（OM-75的细节）
Metal fascia bands suggesting a classical entablature, broken at the division between bays (detail of OM-75)

OM-79 贴面墙系统的金属线脚像夸张的凸圆线脚，当它延伸到拐角时变成扁平状
Metal molding in a panelized wall system resembling an exaggerated bead molding flattened when it turns the corner

OM-80 石基座带有大凸圆线脚和顶边削边的柱基
Stone base molding with a large bead and plinth with a chamfered top edge

OM-81 由齿状饰带组成的抽象的上楣
Abstracted cornice with a suggestion of a dentil band

OM-82 有多种线脚的齿状上楣；窗上方有窗耳（左）
Dentilated cornice with enriched moldings; crossette over window (left)

OM-83 从顶部到底部分别是：托饰带；有希腊回纹饰图案的中楣；有叶形装饰的阿拉伯柱冠；带状柱颈；粗石柱墩
Top to bottom: modillion band; frieze with Greek-key pattern; foliated arabesque capital; banded neck; rusticated corner pier

OM-84 陶立克式柱顶线盘和柱间壁上饰有圆盘以及三竖线花纹装饰的中楣
Doric entablature with block modillion band and frieze with discs in metopes alternating with triglyphs

OM-85 从顶部到底部：心锚形葱形饰、卵锚形圆凸线角、简单饰带、树叶、心形线脚
Top to bottom: heart-and-dart ogee, egg-and-dart ovolo moldings, plain frieze, leaf-and-dart moldings

线脚

OM-86 带有圆锥饰的三竖线花纹装饰和普通墙面交替出现；拱腹上带有圆锥饰的陶立克式檐饰

Triglyphs with guttae alternating with plain metopes; mutules with guttae in the soffit

OM-87 装饰支撑隔带，从顶部到底部：哥特式上楣、三行错齿式线脚、锯齿形饰带、哥特式上楣

Ornamented corbelled courses. Top to bottom: Gothic cornice, three rows of billet molding, zigzag band, Gothic cornice

OM-88 上楣，从顶部到底部：蜿蜒的饰带；托饰饰带；哥特式上楣；底部是齿状拱门饰

Cornice. Top to bottom: serpentine band; modillion band; Gothic cornice. Dentilated archivolt at bottom

OM-89 圆形玫瑰窗上的彩色四叶花图案，反向S形曲线上饰有莨苕叶形装饰图案

Polychrome quatrefoil motif on a circular rose window, with acanthus motif on reverse ogee

OM-90 从顶部到底部：科林斯式上楣；带有玫瑰花图案的花格镶板中间的块状托饰；卵叶形和卵锚形的线脚、简式横饰带、凸圆线脚、舌锚形线脚

Top to bottom: Corinthian cornice; block modillions between coffers with rosettes; egg- and leaf-and-dart moldings, simple fascia, bead-and-reel molding, tongue-and-dart molding

OM-91 有简单托饰的上楣

Cornice with simple modillions

线脚

136

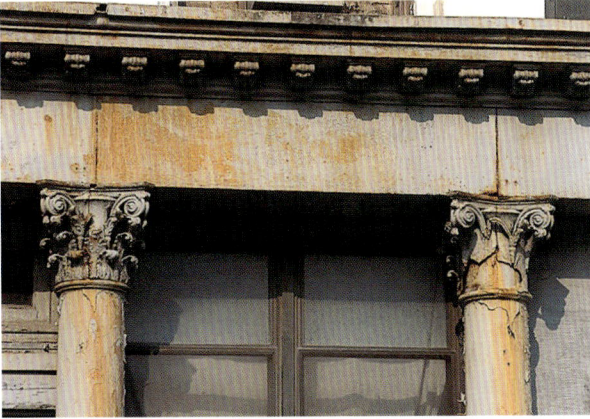

OM-92 古典柱式顶部：简式梁柱过梁，没有中楣风化锡钣外立面上楣的托饰饰带；风化锡钣科林斯式柱冠

Classical entablature: plain architrave, no frieze, modillion band on cornice on weathered tin facade. Weathered tin Corinthian capitals

OM-93 科林斯式古典柱式顶部的拐角细部：简式中楣和线脚；在托饰镶板和拱腹之间的玫瑰图案

Corner detail of Corinthian entablature: plain frieze and moldings; rosettes in coffers and in the soffit between the modillions

OM-94 带有雕刻的和题字的横梁，飞檐托饰带，山墙的三角面部分带有人物雕刻

Architrave with sculpted and inscribed frieze, modillion band; figurative sculpture in tympanum

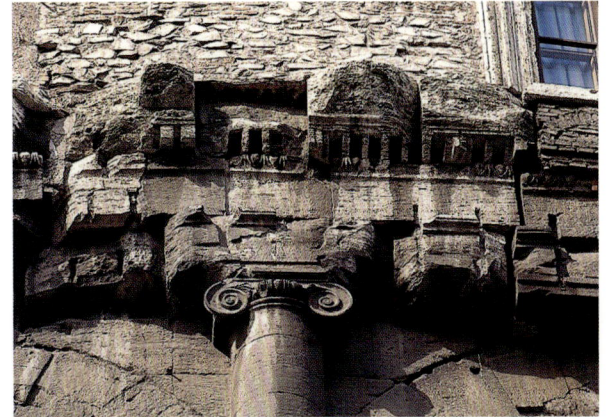

OM-95 饰有残余的部分齿状饰带和一个卵锚形圆凸线脚的风化爱奥尼亚古典柱式顶部

Weathered Ionic entablature with remains of a dentil band and an egg-and-dart ovolo molding below it

OM-96 陶立克式古典柱式顶部：中楣中的花环替代了两个三竖线花纹装饰；拱腹的简式多利安式檐饰

Doric entablature: wreaths substituted for two triglyphs in the frieze; plain mutules in the soffit

OM-97 托饰带下的垂直凹槽饰带，位于壁柱冠颈部

Vertical reeded band below modillion band and on neck of pilaster capital

檐口

OM-98 中楣上有内置嵌板的风化科林斯式上楣
Weathered Corinthian cornice with inset panels in the frieze

OM-99 两个上楣，较低的一个由带有荷叶边齿状图案的简式拱墩的扁平壁柱支撑；中楣的位置上带有栏杆
Two cornices, the lower supported by flat pilasters with molded imposts with scalloped dentil motif; balustrade in the frieze position

OM-100 简式中楣上饰有类似藤蔓纽索饰的科林斯式古典柱式顶部；传统的卷轴石雕侧面式样的托饰
Corinthian entablature with a vine-like guilloche in an otherwise plain frieze; modillions in traditional scroll profile

檐 口

OM-101 上楣有连续人物浮雕的带状装饰
Frieze with a continuous figurative relief on the cornice

OM-102 有陶立克式特征的古典柱式顶部：中楣上的三竖线花纹与纹章装饰交替出现，类似齿状的饰带和没有檐饰的拱腹，典型的科林斯式柱冠
Entablature with Doric features: triglyphs alternating with heraldic ornament in the frieze, quasi-dentil band, and soffit without mutules. Stylized Corinthian capital

OM-103 科林斯式古典柱式顶部，从顶部到底部：挑檐滴水板上的块状托饰与玫瑰花图案交替出现；简式中楣
Corinthian entablature. Top to bottom: block modillions alternating with rosettes in corona soffit; plain frieze

OM-104 双带横梁，被花和彩带图案的带状装饰横饰围绕的科林斯式壁柱
Two-banded architrave, festooned frieze flanking Corinthian pilasters

檐口

摄影: Peter Miller

OM-105 罗马陶立克式风格的古典柱式顶部；墙面中的圆盘装饰；平顶镶板中的玫瑰花图案同拱腹上的简式檐饰交替出现

Roman Doric–style entablature; discs in metopes; rosettes in coffers alternating with plain mutules in the soffit

摄影: Peter Miller

OM-106 三带横梁，简式浅拱腹；横梁上方的突出嵌板被叶锚形线脚环绕

Three-banded architrave, plain shallow soffit; raised panels surrounded by leaf-and-dart molding over architrave

OM-107 带有玫瑰花图案的双带横梁，叶锚形线脚，带花状平纹的凹弧线脚上楣

Two-banded architrave with rosettes, leaf-and-dart molding, cavetto cornice with anthemions

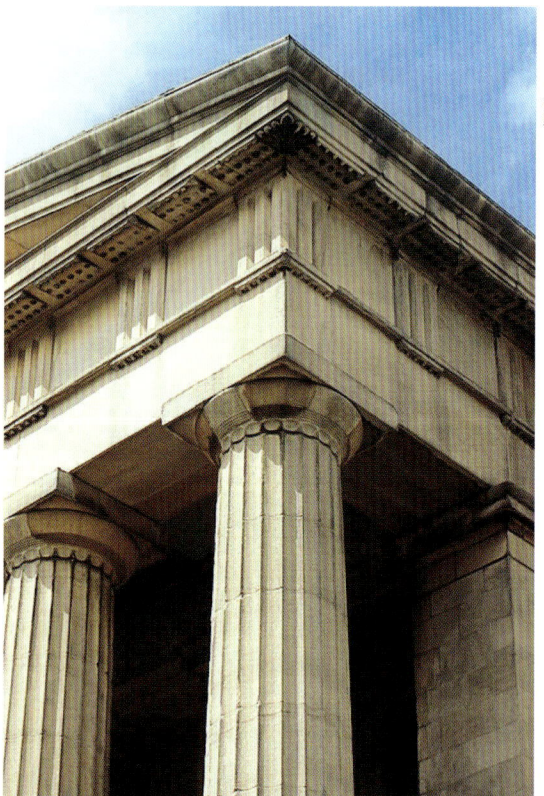

摄影: Peter Miller

OM-108 希腊陶立克式风格的古典柱式顶部：简式横梁，中楣上有三竖线花纹和普通墙面，拱腹上有圆锥饰的檐饰

Greek Doric–style entablature: plain architrave, triglyphs and plain metopes in the frieze, mutules with guttae in the soffit

檐口

柱子、支柱和拱券

摄影：Austin, Patterson, Disston

CP-1 希腊陶立克风格拉毛水泥面支柱，建于1920年
Column c. 1920 in Greek Doric style with a smooth stucco finish

摄影：Peter Miller

CP-2 爱奥尼亚柱冠和希腊风格的古典柱式顶部
Ionic capital and entablature with Greek styling

CP-3 罗马式的陶立克风格柱冠，凸圆线脚中的卵锚图案，
附墙圆柱颈部有玫瑰花雕饰
Doric capital with Roman styling, egg-and-dart motif in echinus;
rosettes on neck of engaged column

CP-4 罗马式爱奥尼亚柱冠
Ionic capital with Roman styling

CP-5 螺旋图案中心有花和彩带装饰的爱奥尼亚柱冠与冠板
上的装饰
Ionic capital with festoon between the eyes of the volutes and
ornament on abacus

CP-6 颈部装饰有玫瑰花图案的有角柱冠与冠板上的花和彩
带装饰
Angular capital with rosette motif on necking and festoon on abacus

柱头

CP-7 罗马风格或拜占庭风格的科林斯式柱冠的配饰
Adaptation of a Corinthian capital in Romanesque or Byzantine style

CP-8 带有中间由半圆饰和莨苕叶形雕饰的独特风格柱颈的混合柱冠
Composite capital with idiosyncratic necking between astragal and acanthus leaves

摄影：Peter Miller

CP-9 科林斯式柱冠和古典柱式顶部
Corinthian capital and entablature

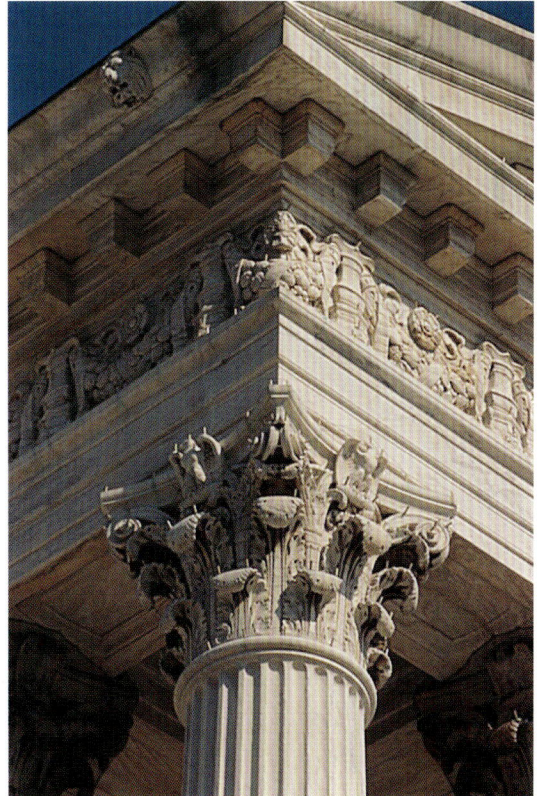

CP-10 科林斯式柱冠和古典柱式顶部
Corinthian capital and entablature

摄影：Peter Miller

柱头

CP-11　有罗马比例的陶立克式柱冠；有特点的希腊式凹槽代替了半圆饰线脚
Doric capital with Roman proportions; characteristic Greek groove replacing astragal molding

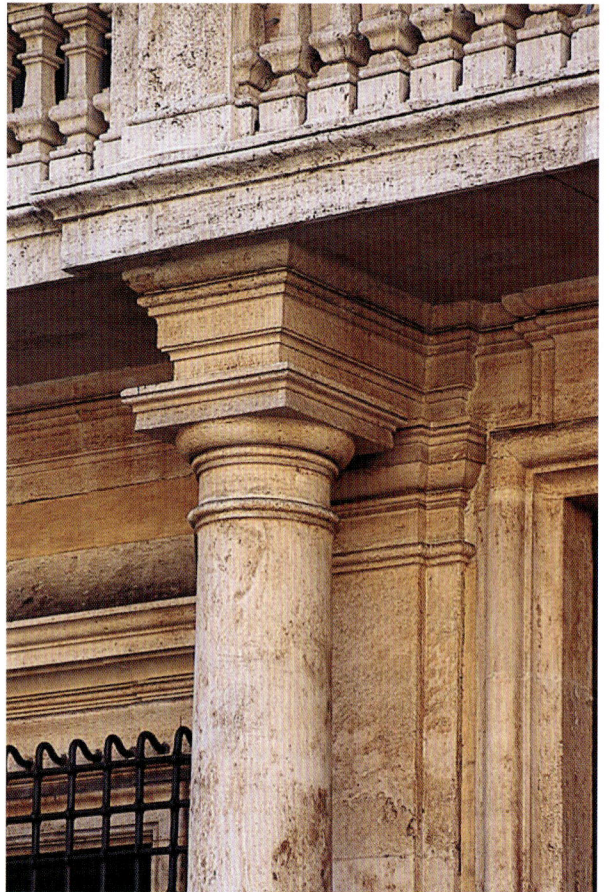

CP-12　有栏杆的悬突体下方的托斯卡纳式风格的柱冠
Tuscan capital under an overhang with balustrade

CP-13　科林斯式柱冠向罗马风格或拜占庭风格的同化
Romanesque- or Byzantine-style adaptation of a Corinthian capital

柱头

CP-14 拜占庭风格方柱冠的现代衍生，用作八边形柱身的拱墩

Modern derivatives of Byzantine-style square capitals serving as imposts on octagonal shafts

CP-15 早期法式哥特风格柱冠的现代衍生，带有超大拱墩和莨苕叶图案

Modern derivative of early French Gothic-style capital with acanthus motif and oversized impost

CP-16 有莲花和莨苕叶形雕饰的柱冠，古希腊风格的衍生

Capital with lotus and acanthus leaves, derivative of ancient Greek style

CP-17 拐角壁柱上的托斯卡纳式风格的柱冠

Tuscan capital on corner pilaster

柱头

CP-18 罗马风格的陶立克风格柱头，邻近附墙圆柱和附壁柱，卵锚形凸圆线脚和典型的陶立克式檐口
Roman-style Doric capitals on adjacent engaged column and pilaster, egg-and-dart echinus, and typical Doric entablature

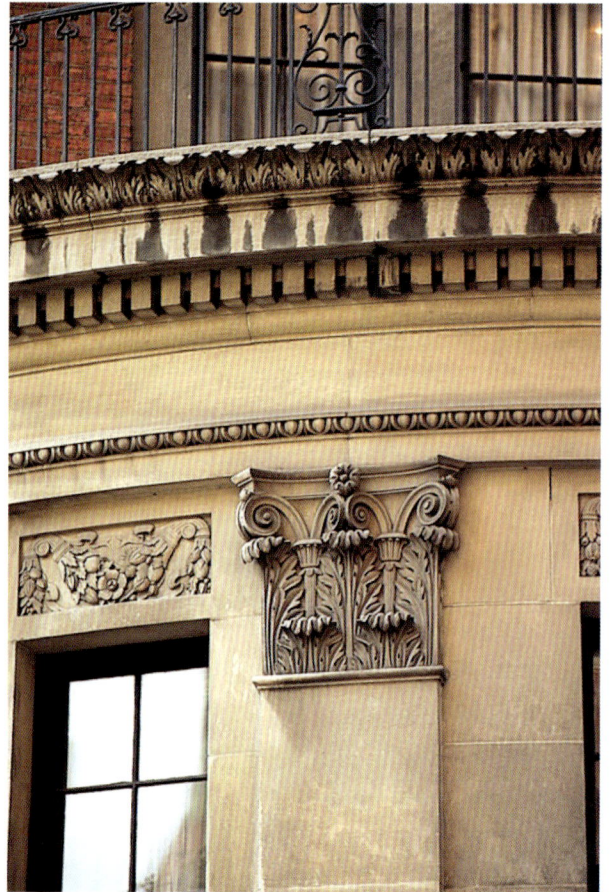

CP-19 有卵锚形线脚和齿状线脚的风化檐口下的带有扁平壁柱的科林斯式柱头
Corinthian capital on flat pilaster below weathered entablature with egg-and-dart and dentilated moldings

CP-20 一列壁柱上的罗马风格柱头和混合拱中的附壁柱
Romanesque capitals on a series of pilasters and jamb-shafts in compound arch

柱头

CP-21 柱头的侧面是陶立克式柱头的衍生；带有齿状线脚的方形角柱石上装饰有花状平纹柱颈
Capital with profile derivative of Doric capital; necking ornamented with anthemion on a square corner pier with dentilated molding

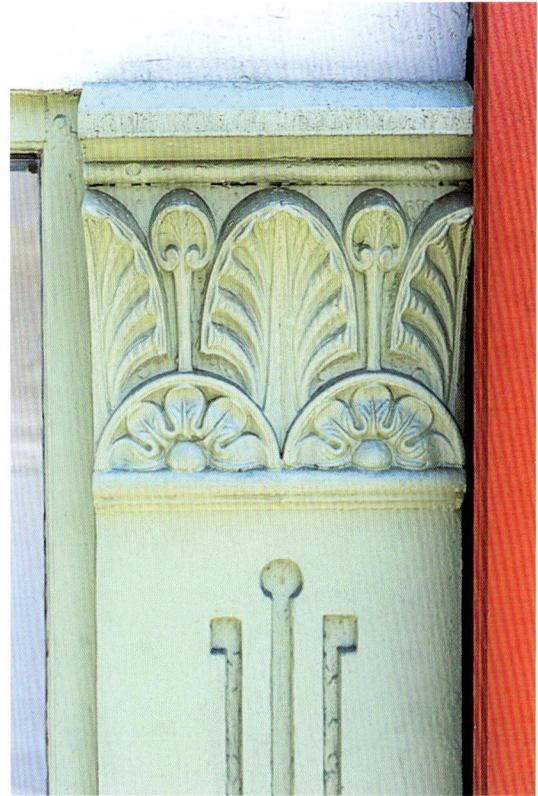

CP-22 铸铁附壁柱和带有莲花图案的柱头
Cast-iron pilaster and capital with lotus motif

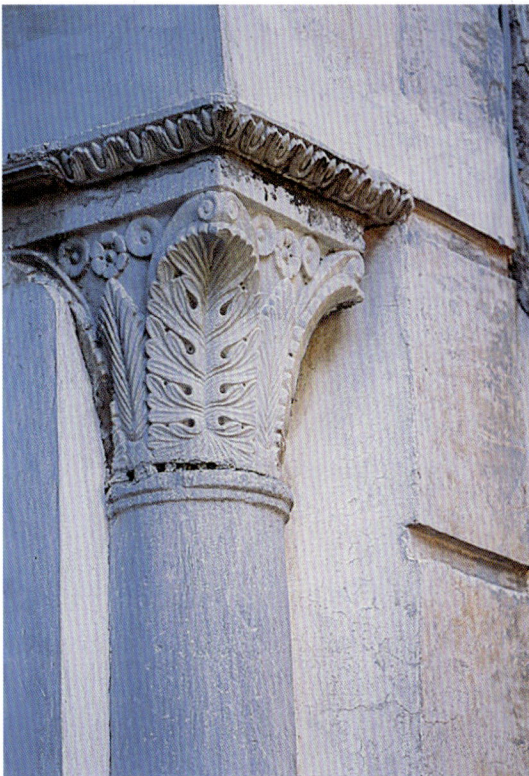

CP-23 入口处的平滑的门窗立柱，拜占庭或埃及风格的柱头
Byzantine- or Egyptian-style capital on smooth jamb-shaft in portal

CP-24 美式柱头和有卵锚形线脚的檐口，阿拉伯式中楣和块状飞檐托饰
American-order capital and entablature with egg-and-dart molding, arabesque frieze, and block modillions

柱头

CP-25　附壁柱上带有凹槽柱颈的爱奥尼亚式柱头
Ionic capital with fluted necking on pilaster

CP-26　墙角带有凹槽柱颈的特殊角度的柱头
Idiosyncratic angle capitals with fluted necking on corner pilaster

CP-27　古老的柱头
Ancient capital

摄影：Guy Gurney

柱头

CP-28　有凹凸条纹的现代方柱头
Modern square capital with fluting

CP-29　石头贴面板对柱头抽象几何式的诠释，现代商业大厦
Abstract geometric reference to a capital in stone veneer, modern commercial building

CP-30　有不规则八边侧面的玻璃和金属材质的装饰艺术风格柱冠，内部有灯光
Art Deco-style capital in glass and metal with non-regular octagonal profile, lit from inside

柱头

CP-31　圆环面柱基覆盖有叶形装饰的古典式大理石基座
Idiosyncratic marble Attic base with foliated ornament draped over torus on plinth

CP-32　有带状突出竖条的火山岩风化支柱的古典式柱基
Attic base of weathered column in volcanic stone with banded rusticated shaft

CP-33　柱基上部圆环面上有叶形装饰的大理石古典基座，柱基与带希腊回纹饰线脚形成对比，并饰有装饰槽
Marble Attic base with foliated ornament on upper torus on plinth in contrasting stone with Greek-key molding; ornamented fluting

CP-34　薄柱基上的古典式大理石基座，从顶部到底部：柱脚圆盘线脚、平线脚、凹形边饰、平线脚、柱脚圆盘线脚
Attic base on a thin plinth. Top to bottom: torus; fillet; scotia; fillet; torus

柱子、支柱和拱券

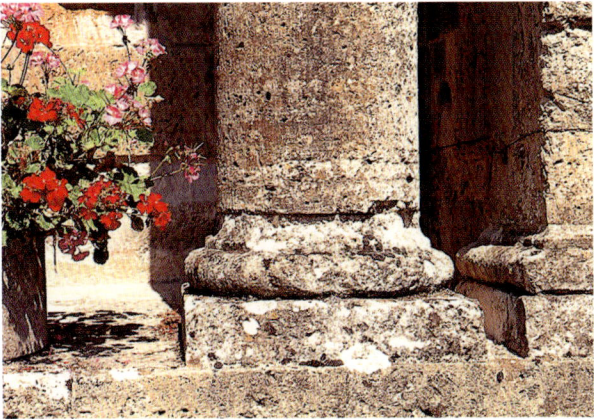

CP-35　方形基座严重风化的火山岩支柱的陶立克式柱基
Doric base of heavily weathered volcanic stone column on square plinth

CP-36　方形基座上风化石灰华柱的古典柱基
Attic base of weathered travertine column on square plinth

CP-37　六边形基座上石灰华附墙圆柱的古典柱基
Attic base of travertine engaged column on octagonal plinth

CP-38　有凹槽细部的方形基座上的古典柱基
Attic base on square plinth with detail of fluting

CP-39　梯状八边形基座上有古典柱基的风化大理石柱，右侧是方形大理石柱基上的砖砌附壁柱
Weathered marble column with Attic base on stepped octagonal plinth; right: brick pilaster on square marble base

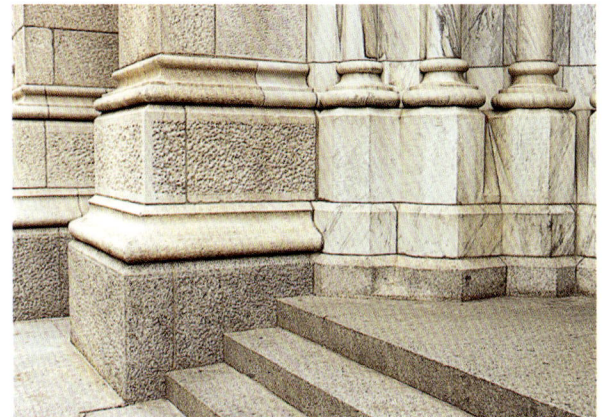

CP-40　左侧：古典外形的大理石墩状柱基和基座下的S形线脚；右侧基座上有古典外形柱基的柱身
Left: marble pier base with Attic profile and ogee molding below plinth; right: jamb-shafts with Attic-profile bases above plinths

基座和柱脚

CP-41 中世纪柱体和火山岩基座上的柱基
Medieval column and base on volcanic stone plinth

CP-42 带状砖石和石头柱身；带有浅的延长的凹形边饰的柱基在扇形边缘墙角的基座之上，基座的柱脚有圆盘线脚
Banded brick and stone jamb-shafts; base with shallow, elongated scotia over torus on plinth with scalloped corner

CP-43 有典型罗马科林斯式爱奥尼亚式基座的方形附壁柱；外角为斜面接合
Square pilaster with typical Roman Corinthian Ionic base; mason's miter on outside corner

CP-44 带有梯状的方形基座的古典柱基上的石灰华柱
Travertine column on Attic base on square stepped plinth

基座和柱脚

CP-45 托斯卡纳式柱基上的带状花岗岩贴面附壁柱，现代商业建筑入口处
Banded granite veneer pilaster on Tuscan-style base, entry of modern commercial building

CP-46 方墩位于带有扇形边缘柱基的几何抽象花岗岩贴面的基座之上，现代商业大厦的门廊
Square pier on base of granite veneer with geometric abstraction of a scalloped plinth, portico of modern commercial building

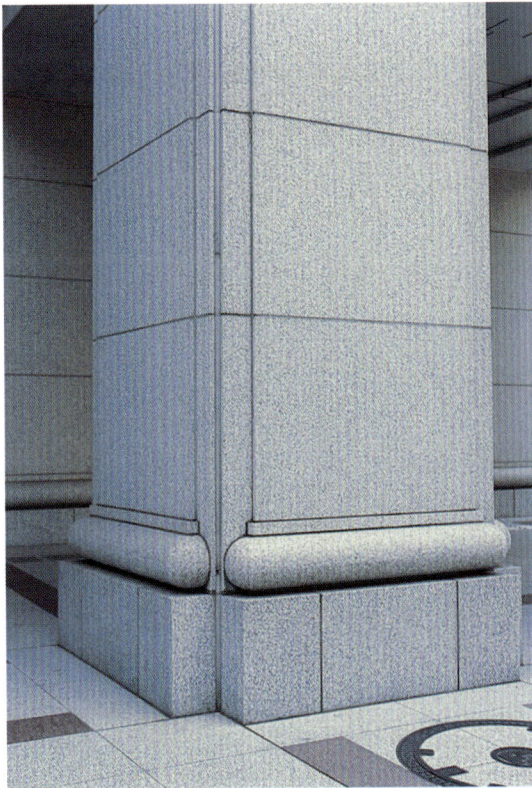

CP-47 托斯卡纳式基座上的花岗岩贴面柱墩，现代商业建筑的入口
Pier of granite veneer on Tuscan-style base, entry of modern commercial building

CP-48 柱基上是带状石板贴面柱墩，墙角和接合处为灌浆的，现代商业建筑
Banded stone-veneer pier on base, grouted corner and joints, modern commercial building

基座和柱脚

CP-49 圆边光面石基座上的手工劈砍的斜面木支柱
Hand-hewn beveled timber post on a honed stone base with rounded edge

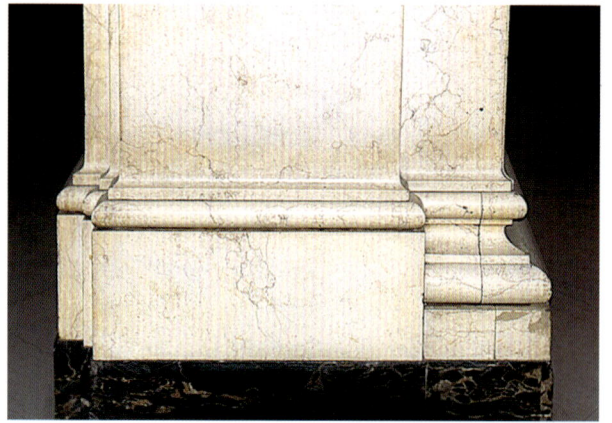

CP-50 复合方形柱基的大理石柱，左侧：高基座上的古典外形柱基；右侧：古典的外形
Marble column with compound square base. Left: Attic profile base on a high plinth; right: Attic profile

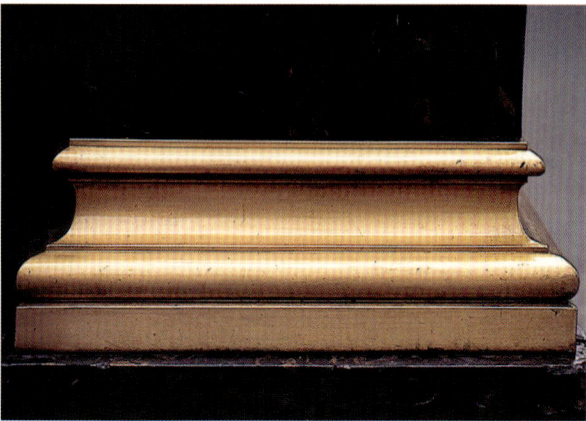

CP-51 方形铁基座，有扩大的凹形边饰和渐弱式上层柱脚圆盘线脚
Square metal base, Attic profile with enlarged scotia and diminished upper torus

CP-52 方形削边石基座上的木支柱
Timber post on a square chamfered stone base

CP-53 日式花园池塘中天然巨砾支撑着凉亭下的木支柱
Natural boulders supporting timber posts under deck of pavilion in Japanese garden pond

基座和柱脚

CP-54 由镀金浮雕装饰金属鞘包裹的木支柱的基座
Gilded relief ornamental metal sheath covering the base of a timber post

CP-55 木工哥特式风格的严重风化的喷漆八边形木柱基
Heavily weathered painted octagonal wood base in Carpenter Gothic style

CP-56 带雕饰的方形石基座上的木支柱，印度尼西亚
Timber post on square stone base with carved design, Indonesia

CP-57 饰有动物雕刻的方形石基座上的木支柱，印度尼西亚
Timber post on stone base with carved animal design, Indonesia

基座和柱脚

CP-58 严重风化的大理石古罗马式带有凹槽的柱子，配有古典基座
Heavily weathered marble ancient Roman fluted column with Attic base

CP-59 有古典基座和花饰的圆形基架，上部为带有陶立克式柱基的凹槽柱身
Fluted column with Doric base on round pedestal with Attic base and floral ornament

CP-60 有哥斯马特式镶嵌图案的大理石所罗门式柱身
Marble Salomonic column shaft with Cosmatesque fluting

柱身

CP-61 线垛陶制柱身，左侧：V形图案；右：网状图案
Bundled terracotta shafts. Left: chevron pattern; right: reticulated pattern

CP-62 严重风化的有古典柱基残余的古罗马柱
Heavily weathered ancient Roman columu with remnants of Attic base

CP-63 带有古典基座的圆形基架上有陶立克式或托斯卡纳式基座和一条圆形装饰带
Cast-iron column with Doric/Tuscan base and a band of round ornaments on round pedestal with Attic base

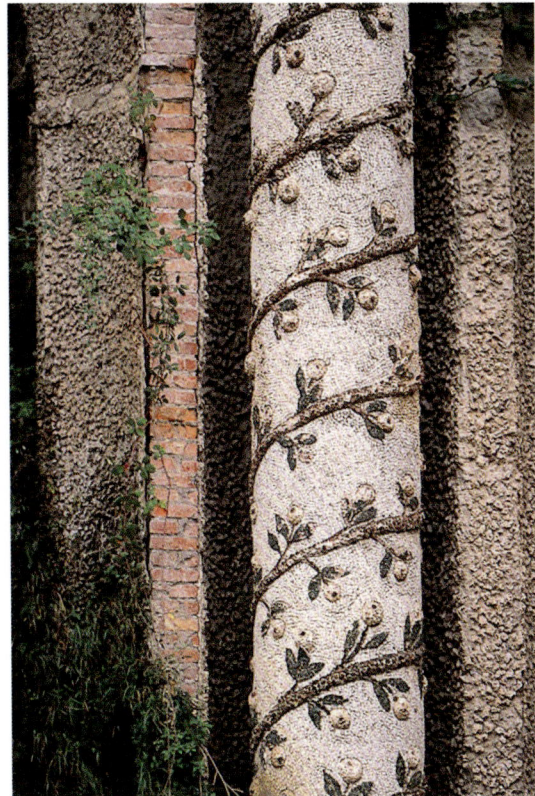

CP-64 饰有马赛克花朵图案的风化柱身
Wreathed column in mosaic with floral design

柱身

CP-65 中间：有花状平纹装饰带的部分有凹槽的柱身和圆柱间的明显接合口；右：有砖块和装饰带的扁平附壁柱
Center: partially fluted shafts with band of anthemion ornament and visible joints between drums; right: flat pilaster with brick and ornamented bands

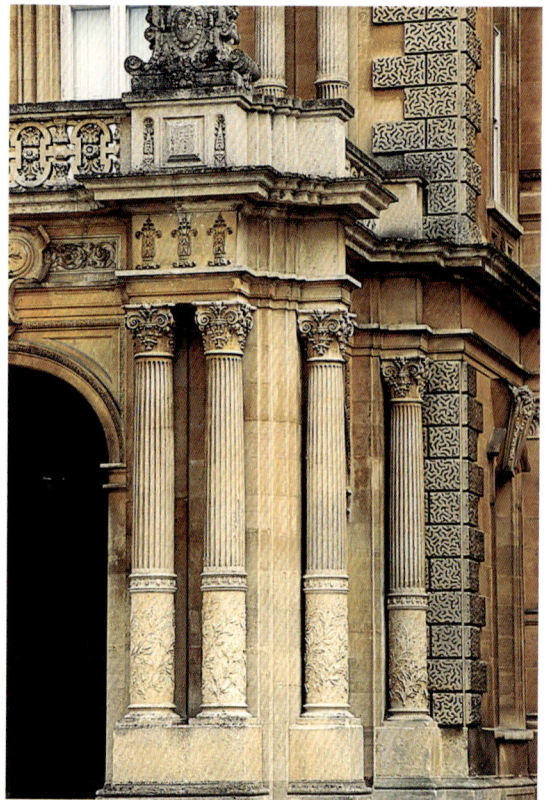

CP-66 有浮雕装饰的部分带有凹槽和带状的科林斯柱
Partially fluted and banded Corinthian columns with relief ornament

CP-67 部分有凹槽的所罗门柱身的石头门廊，带有花环装饰和混合柱头
Stone entryway with partially fluted Salomonic columns, garland ornament, and Composite capitals

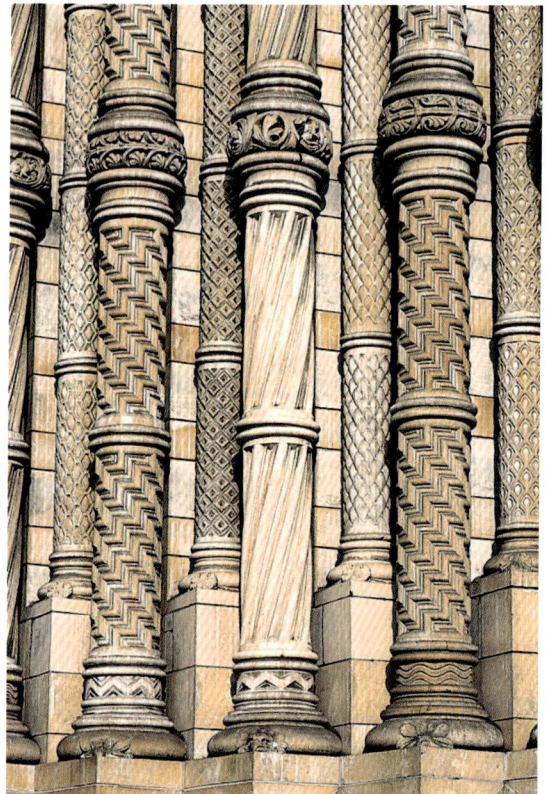

CP-68 带环纹的哥特风格的立柱，有柱身多种图案，包括螺旋形凹槽和梯形几何图案
Gothic-style jamb-shafts with annulets, varied shaft treatments: spiral fluting and stepped helical geometric pattern

柱身

CP-69　有螺旋形凹槽图案的立柱和带螺旋形雕饰的柱头
Jamb-shaft with spiral fluting and angle capital with volutes

CP-70　有凹槽柱身的科林斯式柱子
Corinthian columns with fluted shafts

CP-71　带有未经任何装饰柱头的三根支柱的拐角柱
Corner cluster of three columns with unresolved capitals

CP-72　柱身有部分凹槽的科林斯式圆柱，旁边是扁平的附壁柱
Corinthian columns with partially fluted shafts next to flat pilasters

柱身

CP-73 支撑着门廊的带有古典柱基的罗马风格陶立克式柱，带三角楣饰的门
Roman-style Doric columns with Attic base supporting portico; pedimented door

CP-74 支撑着门廊的科林斯式柱子
Corinthian columns supporting portico

CP-75 柱身有凹槽的大理石爱奥尼亚式柱；有牛眼窗墙；大理石古典檐口
Marble Ionic columns with fluted shafts; wall with bull s-eye windows; marble entablature

柱身

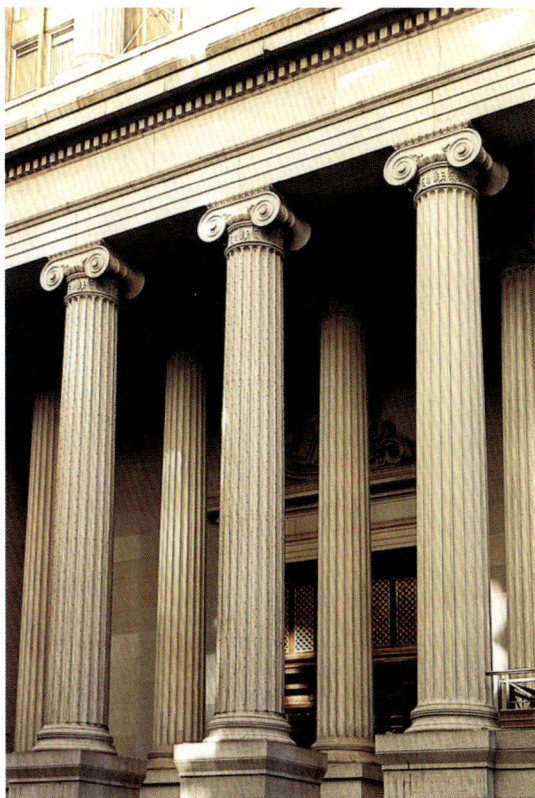

CP-76 带有简洁檐口的双排爱奥尼亚式柱
Double row of Ionic columns with plain entablature

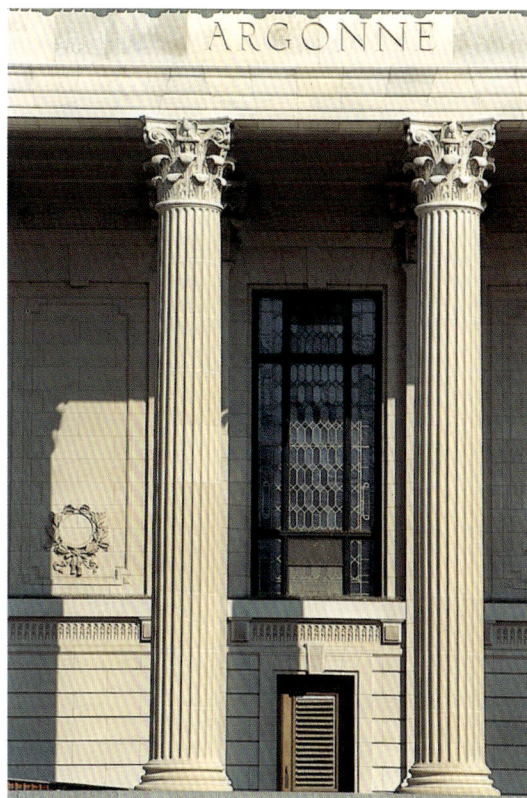

CP-77 科林斯式有凹槽的柱子，檐口上带有文字雕刻
Corinthian fluted columns with engraved entablature

CP-78 明显凸肚状的爱奥尼亚式凹槽柱
Ionic fluted column with pronounced entasis

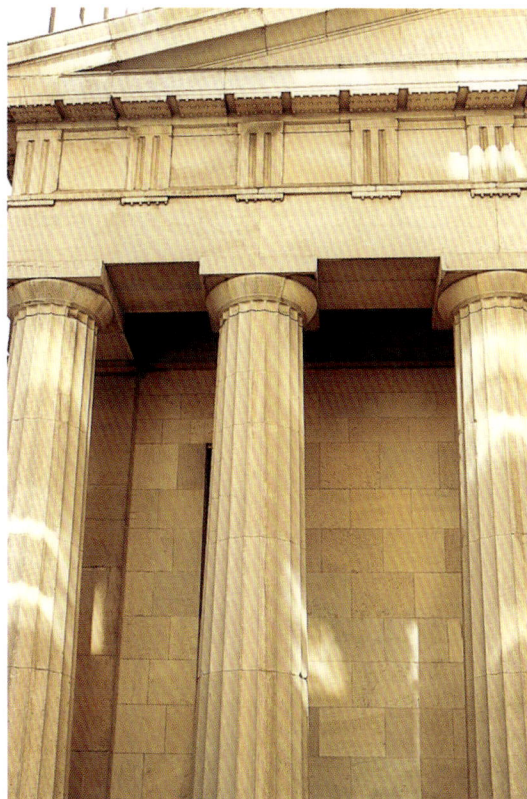

CP-79 有典型檐口的希腊陶立克式柱
Greek Doric columns with typical entablature

柱身

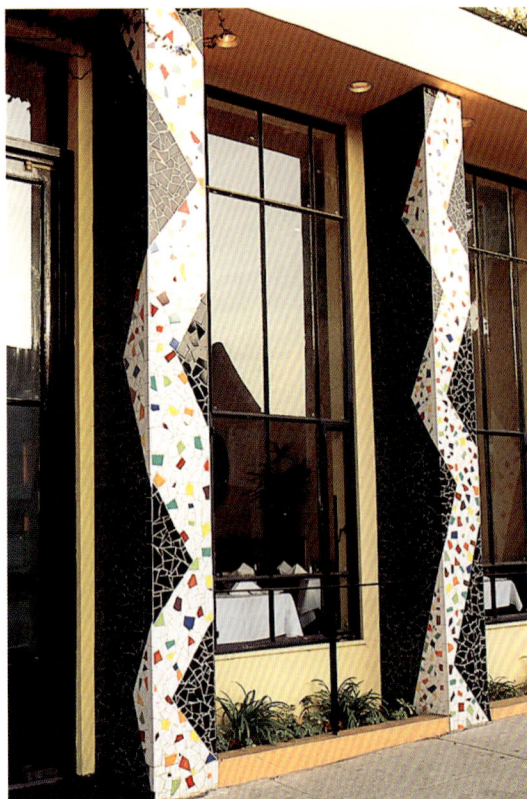

CP-80 饰有"之"字形马赛克图案的附壁柱，餐馆外立面
Pilasters with a mosaic zigzag pattern, restaurant facade

CP-81 混合墙角墩，现代商业建筑
Compound corner pier, modern commercial building

CP-82 带有科林斯风格柱头的齐平凹槽壁柱
Flush fluted pilasters with Corinthian-style capitals

CP-83 相邻的铸铁建筑的壁角柱，左：有凹槽的科林斯式附壁柱；右：有凹槽的带有特别风格的柱头的柱子和与其相临的门窗立柱
Antae of adjacent cast-iron buildings. Left: Corinthian pilaster with fluting; right: fluted shaft with eclectic capital and adjacent jamb-shaft

附壁柱和附墙圆柱

CP-84 有凹槽的石附壁柱和带状粗石柱基
Stone pilasters with flutes and banded rusticated bases

CP-85 突起的蠕虫图案嵌板同砂岩带相交替的带状石附壁柱
Banded stone pilasters with raised vermiculated panels alternating with honed bands

CP-86 光面的爱奥尼亚式建筑陶瓷附壁柱，有角柱冠和有凹槽的柱身；檐口有花和彩带装饰，上部有栏杆
Glazed architectural terracotta Ionic pilasters with fluted shafts and angle-capitals; entablature with festoon, capped with balustrade

CP-87 带梯形侧断面的附壁柱，有卷绳饰和一个类似缩小的埃及钟形柱头
Pilaster with trapezoidal profile, cabled, with a diminished capital resembling an Egyptian bell capital

附壁柱和附墙圆柱

CP-88 带有半圆拱的封闭拱廊中的风化喷漆附壁柱
Weathered painted pilasters in a blind arcade with semicircular arches

CP-89 以有窗口的墙面为背景的带有拜占庭或罗马风格柱头的石附壁柱
Stone pilasters with Byzantine or Romanesque capitals against windowed wall

CP-90 柱头带有齿状图案和封闭拱或很浅的弧形拱饰的砖砌附壁柱；拱耳上方有齿状线脚
Brick pilaster with dentil motif at the capital and blind cambered or very shallow segmental arch; dentilated molding above arch

CP-91 科林斯式扁平附壁柱与带有威尼斯式拱的窗户相邻
Corinthian flat pilaster next to window in a Venetian arch

附壁柱和附墙圆柱

CP-92 科林斯式附墙圆柱
Corinthian engaged columns

CP-93 配有叶形装饰檐口的罗马风格的附墙圆柱三连柱
Triplet of engaged Romanesque columns with foliated entablature

CP-94 粗石附墙带状陶立克式柱和附壁柱；陶立克风格的檐口
Rusticated engaged banded Doric column and pilaster; Doric-style entablature

CP-95 有光面花岗岩柱身的爱奥尼亚式附墙圆柱
Ionic engaged columns with polished granite shafts

附壁柱和附墙圆柱

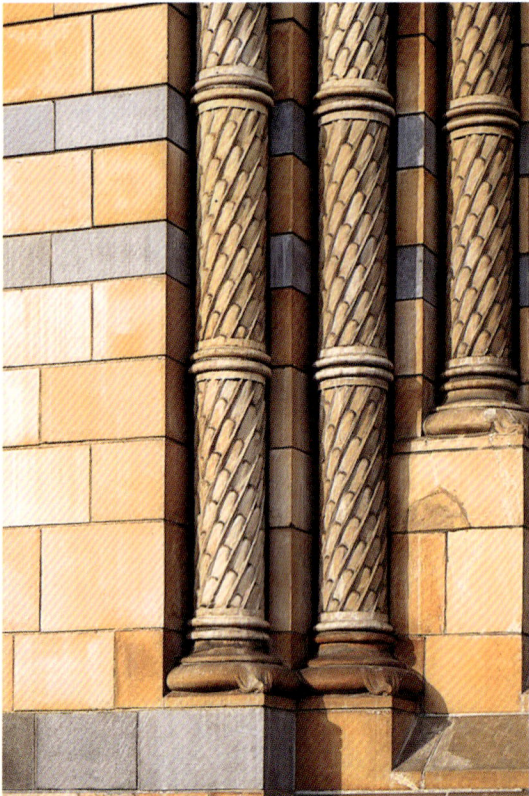

CP-96 入口处饰有环纹的网状附墙圆柱
Reticulated colonnettes (jamb-shafts) in a portal with annulets

CP-97 装饰艺术风格外立面的有凹槽的附壁柱
Fluted and reeded pilasters on Art Deco facade

CP-98 现代商业建筑中的线垛墩柱
Bundled pier on modern commercial building

CP-99 爱奥尼亚式附墙圆柱和装饰性檐口，19世纪美国西南部地区
Engaged Ionic columns and decorated entablature, 19th-c. American Southwestern mission

附壁柱和附墙圆柱

CP-100　罗马风格外立面的线垛式带状墩，现代商业建筑
Bundled banded piers in a Romanesque-style facade, modern commercial building

CP-101　古罗马大理石爱奥尼亚式附墙圆柱
Ancient Roman marble Ionic engaged column

CP-102　楔形砖砌半圆拱形窗两侧为砖砌附墙圆柱
Brick engaged columns flanking gauged-brick semi-circular arch windows

CP-103　带状和砖斜砌的科林斯式附墙圆柱柱头在带有典型楔形石的檐口下方
Banded and rock-pitched engaged Corinthian capitals below entablature with stylized keystones

CP-104　带有檐口和饰有花和彩带装饰拱肩的爱奥尼亚式附墙壁圆柱
Ionic engaged columns with entablature and festooned spandrels

附壁柱和附墙圆柱

CP-105 有榫头和榫眼接合的木柱
Timber post with mortise-and-through-tenon joint

CP-106 木柱用凹槽接合和金属箍连在石头基座上，用榫头和榫眼水平加固
Timber post connected to stone plinth with rabbeted scarf joint and metal hoops. Horizontal brace joined by mortise-and-through-tenon joint

CP-107 梁柱和节点的交叉，梁柱直接穿过一个榫眼并被固定在装饰性的铆铁上
Intersection of post and beam. Post appears to go directly through a mortise and is held in place bydecorative tie irons

CP-108 带有水平支柱的木质梁柱，水平支柱由上部带有金属包条的双楔贯穿榫固定
Timber post with a horizontal brace attached by a double-wedged through-tenon joint with metal stripping on the top

梁柱和节点

CP-109 石墙中的木梁
Timber lintel in stone wall

CP-110 带有两个阴阳榫的木角柱连接于梁柱上
Timber corner post with two offset mortise-and-tenon connections to beams

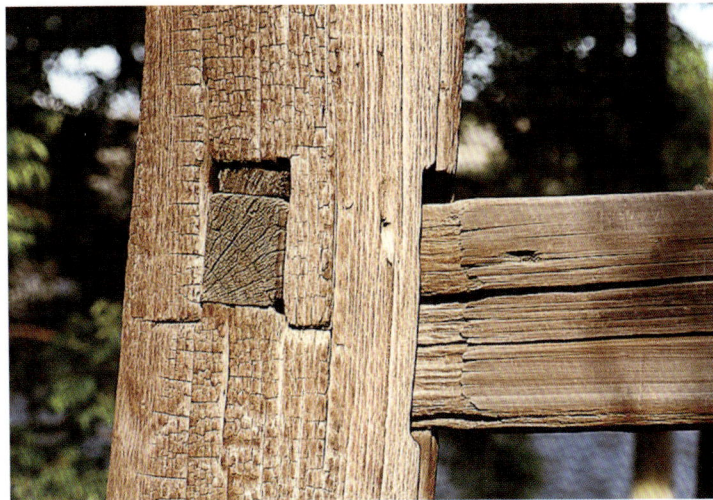

CP-111 梁柱和横梁上相连的阴阳榫接合
Keyed mortise-and-tenon joint between a post and a rail

梁柱和节点

CP-112 钢框架和管状柱；宽缘梁支撑着现代商业建筑的框架玻璃天花
Steel frame and tubular column; wide-flange beams supporting framed glass ceiling of modern commercial building

CP-113 焊接连接的喷漆金属梁；喷漆卷帘式金属墙
Painted metal beam with welded connections; painted corrugated metal walls

CP-114 喷漆现浇混凝土支柱支撑着一个有喷涂拉毛水泥覆层的钢框架
Painted poured-in-place concrete pier supporting a steel frame with cladding finished with spray-on stucco

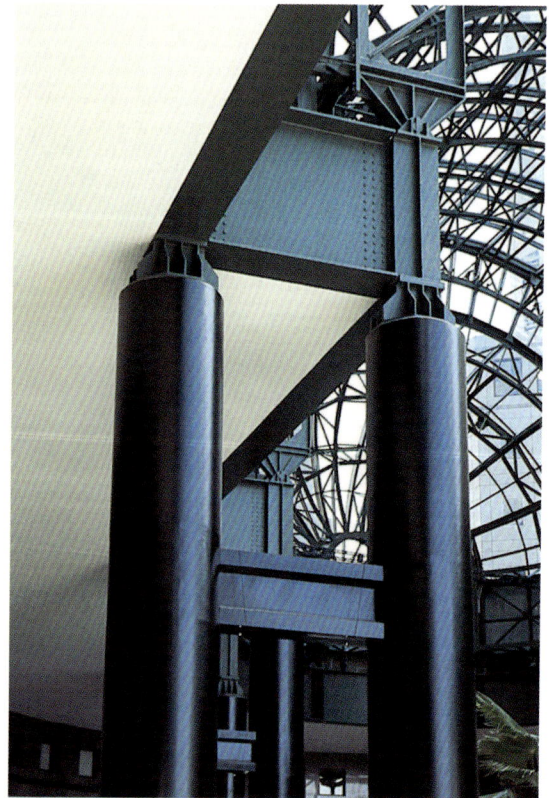

CP-115 现浇混凝土支柱支撑着深色钢板大梁
Poured-in-place concrete piers supporting deep steel plate girders

梁柱和节点

CP-116 悬臂托梁架在支柱上支撑着上一层的梁
Hammer-beam sitting on post supporting a beam above

CP-117 用开裂面石或混凝土建造的支柱，现代商业建筑
Piers built up from split-faced stone or concrete, modern commercial building

CP-118 有焊接和铆接的喷漆金属梁；喷漆波状钢墙
Painted metal beam with welded and riveted connections; painted corrugated metal walls

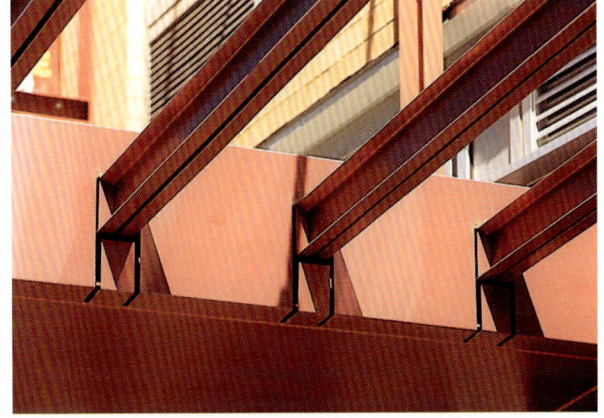

CP-119 用两个C形耐候钢建造的钢梁
Steel beams built up of two C-channels of weathering steel

CP-120 有彩色装饰托架的梁柱结构，日本寺庙
Post-and-beam construction with polychrome ornamental brackets, Japanese templ

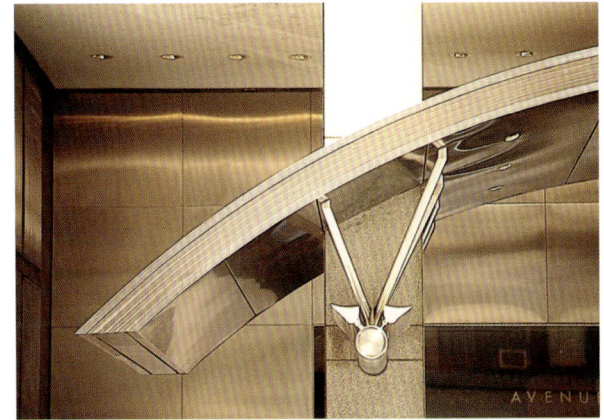

CP-121 由金属托架支撑的弧形金属华盖被固定在砖石柱上，现代商业建筑
Semicircular metal canopy supported by a metal bracket set into a masonry column, modern commercial building

梁柱和节点

摄影：Austin, Patterson, Disston

CP-122 有宽缘梁和铆钉连接的喷漆钢框架，支撑着波状钢
盖板，现代房屋天花
Painted steel frame with wide-flange beams and riveted
connections, supporting corrugated steel decking, ceiling of
modern house

CP-123 带有金属覆层圆柱和用铆钉固定的支架下的梁柱，
现代商业建筑
Detail of steel frame with metal-clad column and beam beneath
exposed riveted plate girders, modern commercial building

CP-124 钢框架里包裹着不锈钢嵌板，现代商业建筑
Steel frame clad with possibly stainless steel panels, modern
commercial building

CP-125 用铆钉固定的钢板上的喷漆钢桁架，工业建筑
Painted steel truss of riveted plates, industrial building

梁柱和节点

CP-126 带有光面石和金属包层的钢支墩或混凝土支墩，现代商业建筑
Steel or concrete piers with polished stone and metal cladding, modern commercial building

CP-127 多边形混凝土支柱支撑着建筑的突出部分，现代商业建筑中的倒影池
Polygonal concrete pier supporting overhang, reflecting pool in modern commercial building

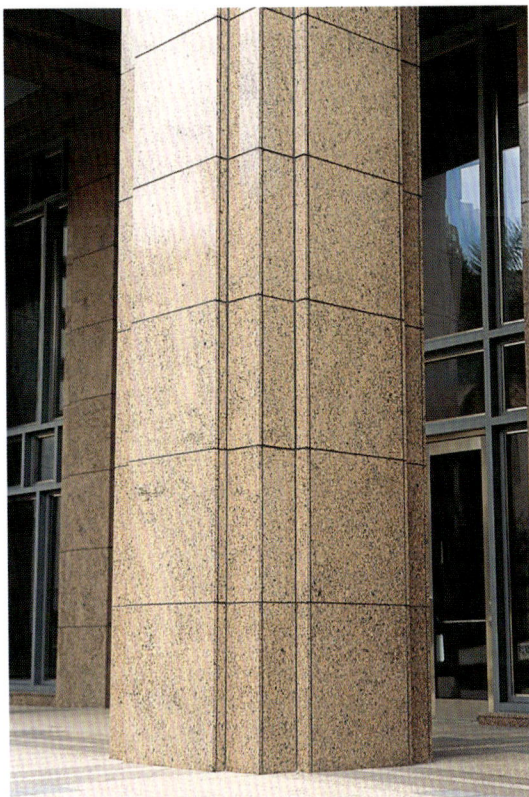

CP-128 拐角处有接合的长方形贴面石板支柱支撑着门廊，现代商业建筑
Rectangular stone-veneer pier with articulated corners supporting portico, modern commercial building

CP-129 不锈钢覆层的钢柱支撑着入口的檐篷，现代商业建筑
Steel column with stainless-steel cladding supporting entry canopy, modern commercial building

梁柱和节点

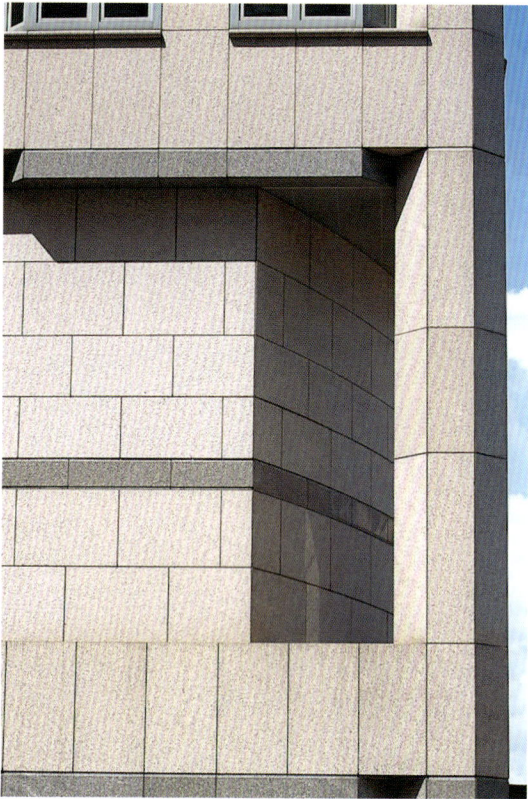

CP-130 带有石质饰面板的墙角八边形支柱
Exposed corner octagonal pier with stone veneer

CP-131 有金属梁柱和石质贴面板的支柱支撑着入口处的檐篷，现代商业建筑
Pier with metal post and stone cladding supporting entry canopy, modern commercial building

CP-132 混凝土支柱和梁柱，可能为预制混凝土，支撑着井式楼板
Concrete column and beams, possibly pre-cast, supporting waffle slab

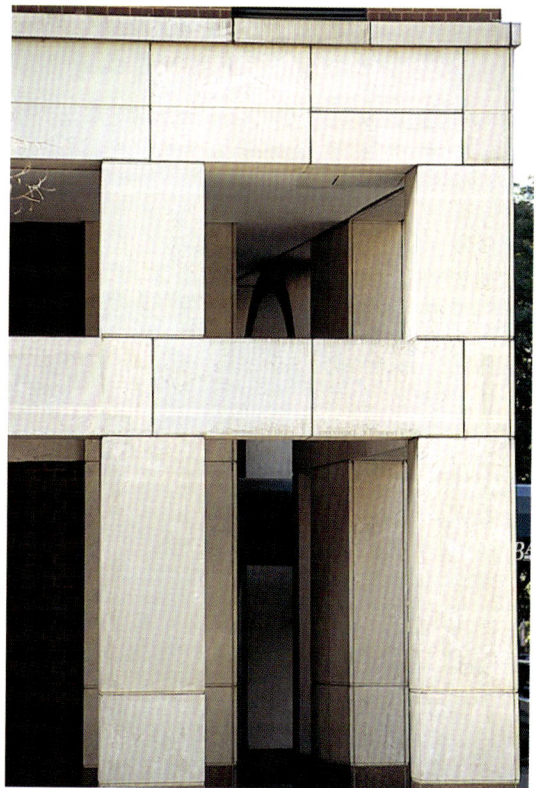

CP-133 有横梁的拱廊（连梁柱结构），现代商业建筑
Trabeated arcade (post-and-lintel construction), modern commercial building

梁柱和节点

CP-134 圆形石座上的木柱支撑着走廊顶部
Timber post on turned stone base supporting roof of veranda

CP-135 方柱上带状蠕虫图案的软质石灰石附壁柱，花园拱廊
Banded vermiculated coquina pilasters on square posts, garden arcade

CP-136 拱廊中有相邻箱式梁的光面粗石花岗岩侧面支柱，现代商业建筑
Polished rusticated granite-veneer piers with adjoining boxed beams in arcade, modern commercial building

CP-137 有古典柱基的石梁柱支撑着花园的木格架
Stone post with Attic base supporting wooden garden trellis

梁柱和节点

CP-138 有金属覆层的钢架和横梁下铆钉固定的板梁，现代商业建筑
Steel frame with metal-clad column and beam beneath exposed riveted plate girders, modern commercial building

CP-139 饰有犬齿隔带的砖支柱
Brick piers with dog-tooth courses

CP-140 梁柱结构，日本寺庙
Post-and-beam construction, Japanese temple

梁柱和节点

CP-141 饰有维特鲁威式卷形装饰和玫瑰花图形圆饰的顺砖砌合墙上的有拱心石的半圆拱
Semicircular arch with an ancon keystone on masonry running-bond wall with Vitruvian scroll trim and rosette medallions

CP-142 带拱石和拱门饰的半圆，雕刻拱肩和齿状线脚
Semicircular arches with voussoirs and archivolt, carved spandrel, and dentilated molding

CP-143 带有模筑拱基石、拱门饰和内置铸铁花饰窗格的拉毛水泥墙上的半圆拱门
Semicircular arch in stucco wall with molded impost blocks, archivolt, and inset wrought-iron tracery

CP-144 弧形混合拱的拱廊；带有螺旋形装饰和叶形装饰的拱肩
Arcade of semicircular compound arches; spandrels with cartouches and foliated ornament

CP-145 以柱顶过梁为延续的拱门饰；门上有三角墙
Archivolt continuous with architrave; pediment over the door

拱券

CP-146 窗四周有装饰性拱心石的弧形拱，有绳状和叶形装饰线脚
Semicircular arch around window with ornamented keystone, rope and foliated moldings

CP-147 有梁托的钟形拱，被有图案的面砖环绕的几何形拱门饰
Bell arch with corbels, geometric archivolt surrounded by patterned terracotta

CP-148 弧形拱的拱廊，内弧面中有绳索形线脚、玫瑰花图案方形嵌板，拱侧上有四叶饰圆形装饰
Arcade of semicircular arches, rope molding in intrados, square panel rosettes, and quatrefoil medallion in spandrel

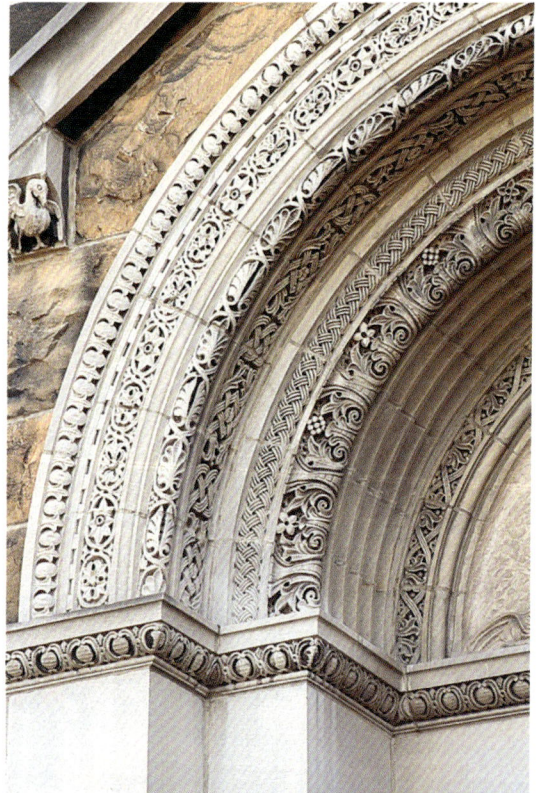

CP-149 有雕刻拱门饰的混合拱
Compound arch with carved archivolts

CP-150 原生岩碎石墙面中的粗面半圆砌石拱
Rustic semicircular masonry arch in rubble wall in natural rock formation

CP-151 带有面砖拱门饰的摩尔式马蹄形拱
Moorish horseshoe arch with terracotta archivolt

CP-152 拱心石上有螺旋形装饰的梯状独立拱
Stepped freestanding arch with cartouche over keystone

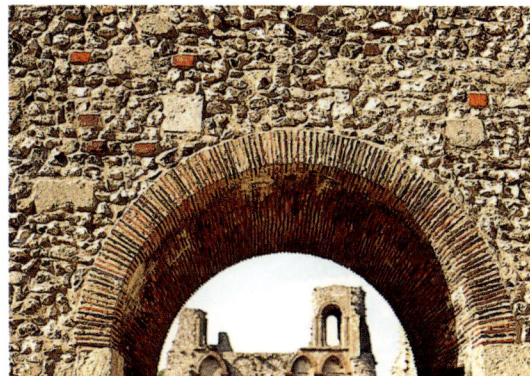

CP-153 随意堆砌的碎石墙中的弧形罗马式砖拱
Semicircular Roman brick arch in random rubble wall

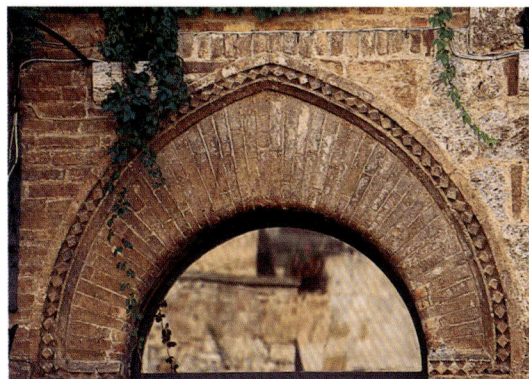

CP-154 佛罗伦萨式拱（拱背和内弧不同中心），楔形砖，被线脚突显的拱背
Florentine arch (extrados and intrados not concentric), gauged brick. Extrados emphasized with a molding

CP-155 城堡入口梁托下的双中心装饰线拱（拱高小于其跨度的一半）
Two-centered surbased arch (arch rising less than half its span) in castle entry under corbelled jetty

拱券

CP–156 面砖墙面和拱楣上带有装饰图案的半圆拱
Terracotta wall and semicircular arch with ornamented tympanum

CP–157 入口上方半圆的楔形砖拱
Semicircular gauged-brick arch over entryway

CP–158 装有木门的砖石砌筑的浅扁园拱
Shallow surbased arch in rubble masonry with wood door

拱券

CP-159 门楣中心有人物浮雕的半圆混合拱
Semicircular compound arch with carved relief sculpture in tympanum

CP-160 装有双扇门的风化拉毛水泥面半圆石拱，乡村庭院
Weathered stucco-covered stone semicircular arch with wood double doors, rustic courtyard

CP-161 半圆拱中的窗户；有雕刻线脚的拱背接在柱头上
Windows in semicircular arch; extrados with carved molding terminating at capitals

拱券

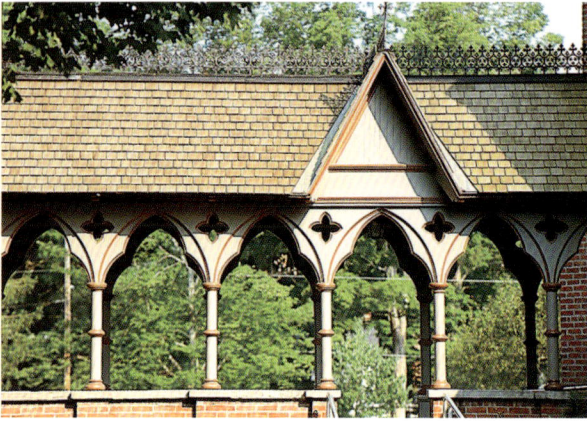

CP-162 有三叶拱的木工哥特式拱廊，拱肩上有四叶草装饰
Carpenter Gothic arcade with trefoil arches, quatrefoil ornaments in spandrels

CP-163 带有伊斯兰风格扁圆拱的拱廊（左侧的三个为葱形拱）；方柱上为拜占庭式柱头
Arcade with Islamic-style surbased arches (the three on left, ogee arches); Byzantine capitals on square columns

CP-164 起拱点有线脚的半圆形拱，风干砖外有拉毛水泥，18世纪晚期美国西南部地区
Arcade with semicircular arches with molding at springing, stucco over adobe brick, late-18th-c. American Southwestern mission

CP-165 弧形拱，砖外层有拉毛水泥，瓦顶
Semicircular arcade, stucco over brick; tile roof

摄影：Guy Gurney

CP-166 有多个拱顶和拱的穹隆拱廊
Groin-vaulted arcade with multipart vaults and arches

拱券

CP-167 三层风化的古罗马式拱
Three levels of weathered ancient Roman arches

CP-168 有弧形梯状拱门、门耳、饰带和饰有怪状石像的粗石贝壳岩花园入口
Rusticated coquina garden portal with semicircular stepped arch, crossettes, banding, and ornamented balustrade with a mascaron

摄影：Guy Gurney

CP-169 带有斜纹拱顶的哥特式拱廊
Gothic arcade with ribbed vaulting

拱券

CP-170 半圆拱包括三个位于六叶草装饰和两个三叶草装饰之下的
复合尖角三叶拱窗口，并带有栏杆
Semicircular arch enclosing three compound pointed trefoil arched
windows below sexfoil and two trefoil ornaments, with balustrade

CP-171 封闭拱：哥特式风格外立面上的垂拱和三叶封闭拱廊
Blind arcade: drop arches and trefoil blind arcade on Gothic-style facade

CP-172 封闭拱：有装饰的石质三心拱，18世纪晚期庭院
Blind arcade: basket-handle (three-centered arch) stone arches with ornament, late-18th-c. courtyard

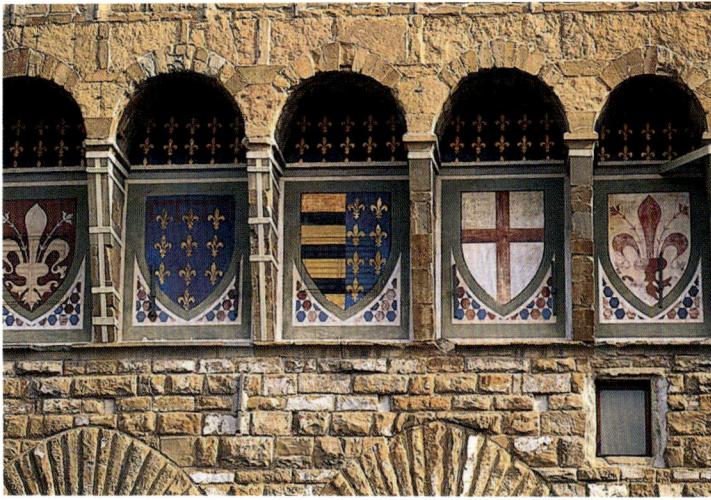

CP-173 带有盾徽的上层弧形拱中的封闭拱
Upper-story blind arcade of semicircular arches with coats-of-arms

CP-174 带窗拱廊，斜面窗口以半圆拱为框
Fenestrated arcade of semicircular arches framing embrasure windows

CP-175 带有花饰窗格的都铎式拱
Arcade of Tudor arches with tracery

拱券

CP-176 圆筒形拱顶，金属和玻璃结构的现代温室
Barrel-vaulted metal and glass modern greenhouse structure

CP-177 圆筒形拱顶的钢管混凝土结构玻璃入口，现代商业建筑
Barrel-vaulted tubular-steel-frame-and-glass entry, modern commercial building

CP-178 多层次的拱廊，现代公寓
Multiple levels of arcades, modern apartment house

拱券

窗户

WI-1　铅条镶嵌圆形玻璃
Leaded roundels

WI-2　石拱上带有装饰的铅条镶嵌玻璃
Leaded window with decorative cames in stone arch

WI-3　菱形格子架图案的铅条镶嵌坚铰链窗
Leaded casement window with cames in diamond lattice pattern

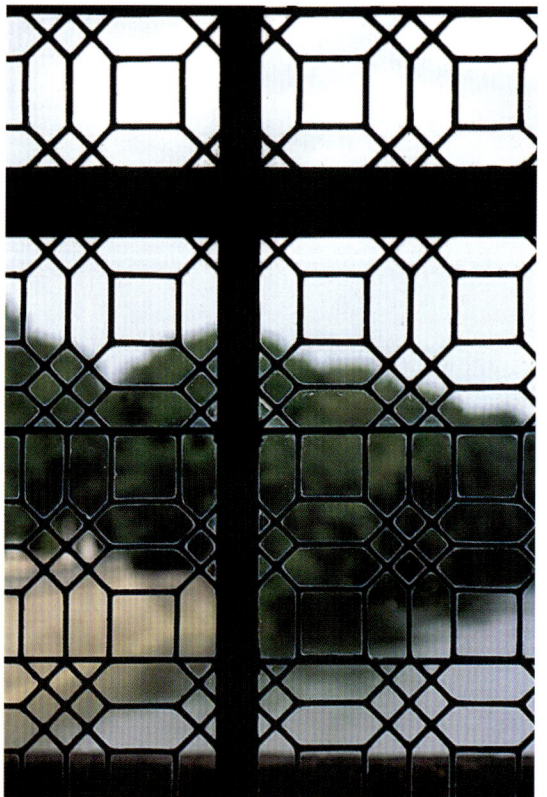

WI-4　套叠图案的八边形黄铜窗格条镶嵌玻璃
Leaded glass with octagonal brass muntins in interlocking pattern

直棂、窗框和镶铅条窗玻璃

WI-5 黄铜色的窗格条和竖棂的固定窗，现代商业建筑
Fixed window with brass muntins and mullion, modern commercial building

WI-6 带有不锈钢窗格条和竖棂的固定窗，现代商业建筑
Fixed window with stainless steel muntins and mullions, modern commercial building

WI-7 不锈钢窗格，现代商业建筑
Stainless steel muntins, modern commercialbuilding

WI-8 金属窗格条，固定着已风化的半透明玻璃，工业建筑
Metal muntins holding weathered translucent glass, industrial building

直棂、窗框和镶铅条窗玻璃

WI-9　铝条镶嵌彩色肌理玻璃，20世纪美国房屋
Leaded colored and textured glass, 20th-c. American house

WI-10　有不同纹理图案的压花玻璃
Figured glass with a pattern of different textures

WI-11　金属框架中带有扁平金属窗格条的压花玻璃，工业建筑
Figured glass with flat metal muntins in a metal frame, industrial building

WI-12　被三个竖梃分开的镶有着色玻璃的窗格条，工业建筑
Muntins with tinted glass separated by three mullions, industrial building

WI-13　风化的着色玻璃，工厂窗户
Tinted and weathered glass, factory windows

WI-14　磨砂玻璃上的透明蚀花设计
Etched design in clear glass on frosted ground

彩色的、有织纹的玻璃

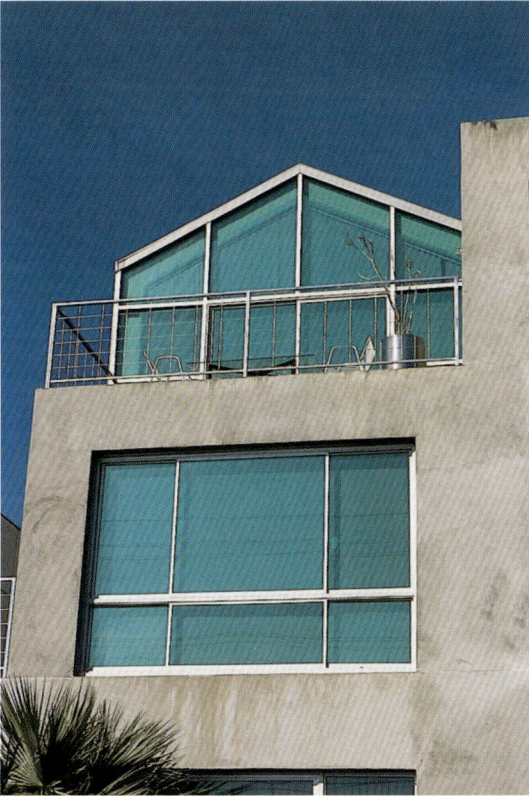

WI-15 下层窗户和山墙的不锈钢框中的彩色玻璃，现代多功能建筑
Colored glass in stainless steel frame in lower window and gable wall, modern mixed-use building

WI-16 彩色的、透明的固定坚铰链窗（注意中间几扇窗户后的楼板），现代房屋
Colored and clear fixed and casement windows (note floor plate behind the middle windows), modern house

WI-17 幕墙上古铜色的镜面玻璃，现代商业建筑
Bronze mirror-glass windows in curtain wall, modern commercial building

WI-18 古代彩色玻璃，19世纪多功能建筑的大厅
Antique colored glass, lobby of 19th-c. mixed-use building

彩色的、有织纹的玻璃

WI-19 玻璃砖的拐角窗口细部
Detail of glass-block corner window

WI-20 透明玻璃和磨砂玻璃的阶梯状窗户，中间带有马赛克嵌板，现代房屋
Stepped window in clear and frosted glass block with ceramic tile central panel, modern house

WI-21 玻璃砖嵌板跨两层楼高，两侧是固定窗，现代商业建筑
Glass-block panels spanning two stories and flanked by fixed windows, modern commercial building

WI-22 波状窗户：平面的和弯曲的波纹玻璃砖，装饰艺术风格公寓建筑
Undulating window: flat and curved glass block with wavy texture, Art Deco apartment building

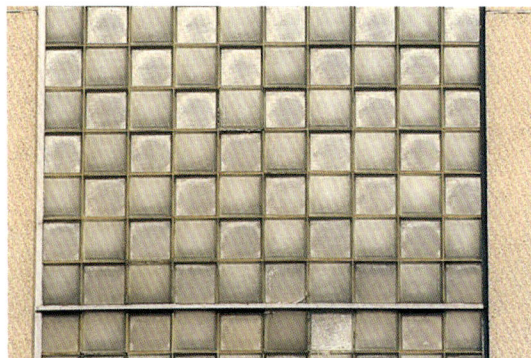

WI-23 棋盘形图案线形纹理玻璃砖，20世纪多功能建筑
Linear textured blocks in checkerboard pattern, 20th-c. mixed-use building

WI-24 嵌有玻璃砖的圆形窗户，现代商业建筑
Circular windows with glass block, modern commercial building

玻璃砖

WI-25　带分开的气窗的两个双开窗，石头窗沿，19世纪公寓
Two double-casement windows with divided transom lights, stone window surrounds, 19th-c. apartment house

WI-26　木框的上悬窗，带有雕刻气窗面板，巴厘岛房屋
Wood-framed awning windows with carved transom panel, Balinese house

WI-27　带有菱形图案的铅条镶嵌玻璃双开窗，16世纪英国半木结构房屋
Leaded double-casement windows with cames in a diamond pattern, 16th-c. half-timbered English house

WI-28　带有造型木窗格条的木框上悬窗，艺术手工房屋
Wood-framed awning windows with shaped wood muntins, Arts and Crafts house

WI-29　窗扇中间有附壁柱的威尼斯风格的双开窗
Venetian-style casement windows with pilasters between windows

窗户种类

WI-30 三扇窗户带有防蚊纱窗的六格或十格窗格的平开窗，19世纪城镇房屋
Six- and ten-light double-casement windows with fly-screen insets in three windows, 19th-c. town house

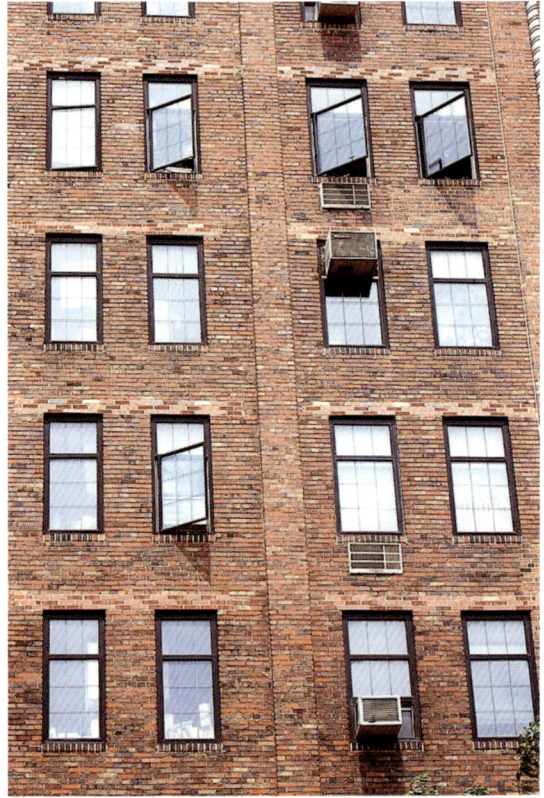

WI-31 有内部窗格条和上层固定气窗的平开窗，老式公寓中替换的现代金属窗
Casement windows with internal muntins and fixed light above, modern metal replacement windows in older apartment house

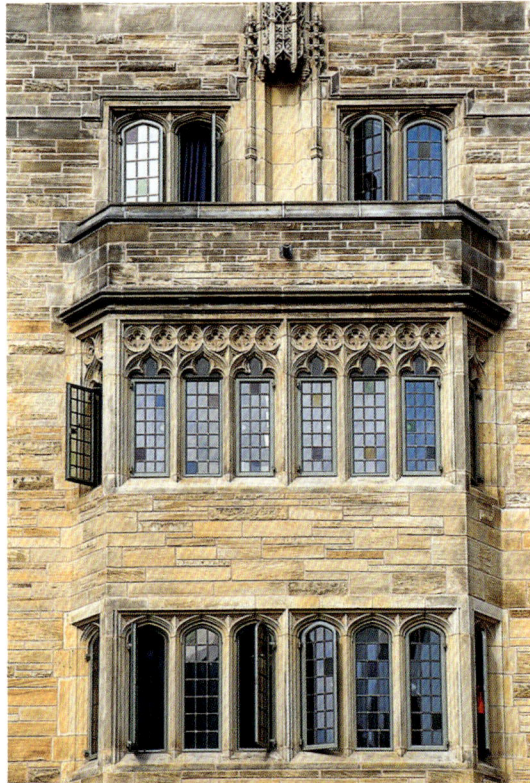

WI-32 新哥特式凸窗中的铝条镶嵌平开窗；中间的窗户有石头花饰窗格的
Leaded casement windows in neo-Gothic bay window; middle windows with stone tracery

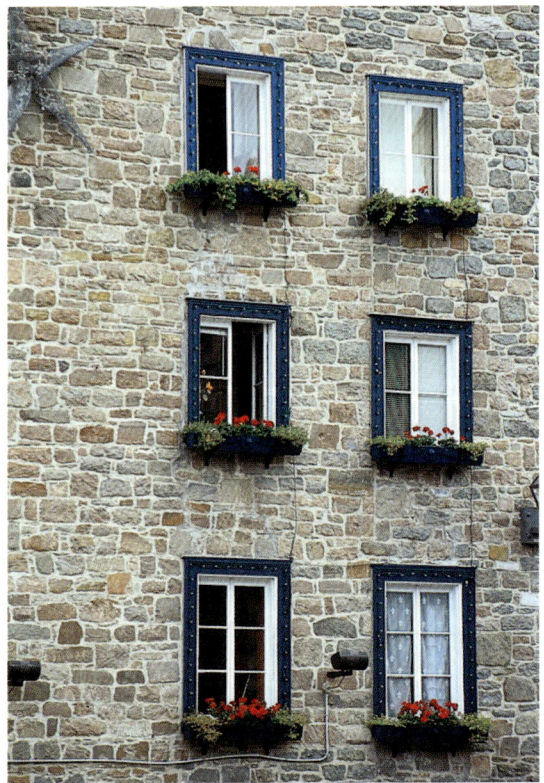

WI-33 两格和三格窗格的双开窗
Two- and three-light double-casement windows

WI-34　铅条镶嵌玻璃的平开窗，上面的窗口在墙壁上方，石砌房屋
Leaded casement windows, top set in wall dormer, stone house

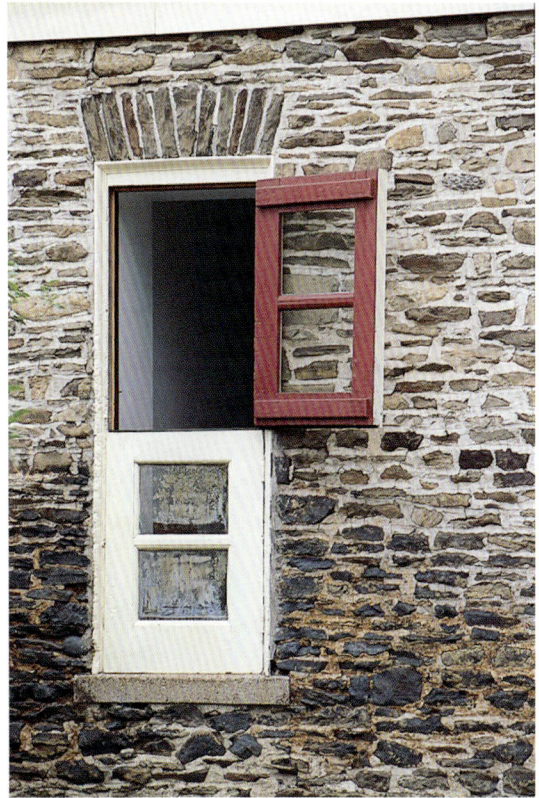

WI-35　石头墙面上的分隔式木框平开窗
Split-casement windows with wood frames in stone wall

WI-36　拉毛水泥墙面上的双开窗
Double-casement window in stucco wall

WI-37　弯曲墙角的翻窗组，装饰艺术风格公寓
Awning windows in curved corner, Art Deco apartment house

窗户种类

WI-38　平开窗或下悬窗安装在类似挤压铝型材外框的固定窗户上，现代房屋
Casement or awning windows set in fixed windows with probably extruded aluminum framing, modern townhouse

WI-39　平开窗和上悬窗（左）安装在固定窗户上；挤压铝型材竖棂，现代公寓
Casement windows (center) and awning window (left) set in fixed lights; extruded aluminum mullions, modern apartment house

WI-40　全上釉木拉门，四周是固定窗或者上悬窗，现代房屋
Fully glazed wood sliding door, surrounded by fixed lights or awning windows, modern house

WI-41　堆叠的上悬窗
Stacked awning windows

WI-42　有不同窗格条图案的单悬窗或上悬窗，现代公寓房屋
Single-hung or awning windows with various muntin patterns, modern apartment house

WI-43　上悬窗，现代金属覆层房屋
Awning windows, modern metal-clad house

窗户种类

WI-44　上下双层双悬窗
One-over-one double-hung windows

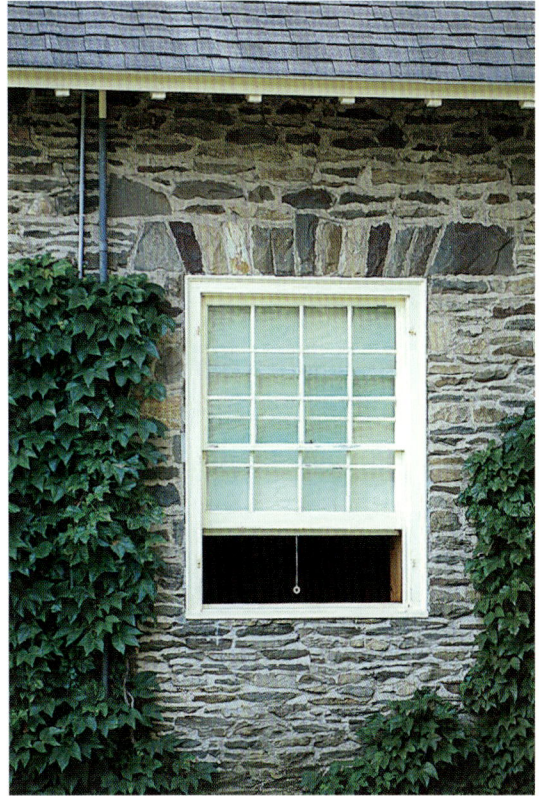

WI-45　上下窗扇都是12个窗格的双悬窗，美国石砌房屋
Twelve-over-twelve double-hung window, American stone house

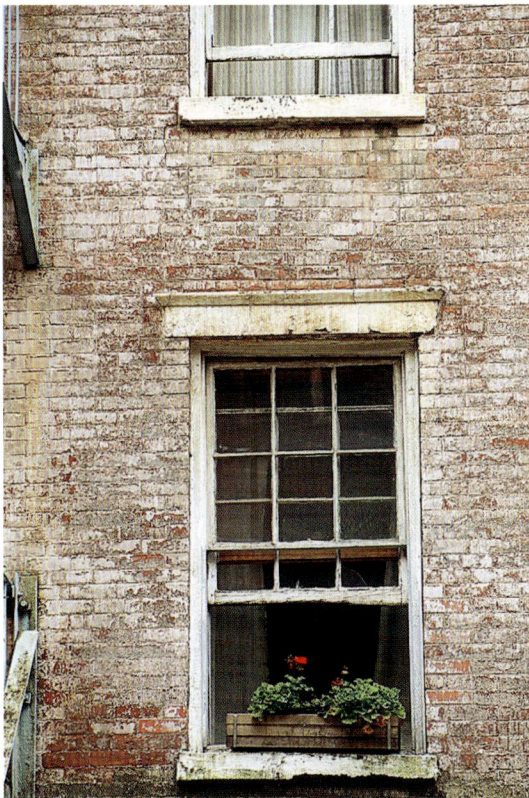

WI-46　风化砖墙中有石头窗楣的六格双悬窗
Six-over-six double-hung window with stone lintel in weathered brick wall

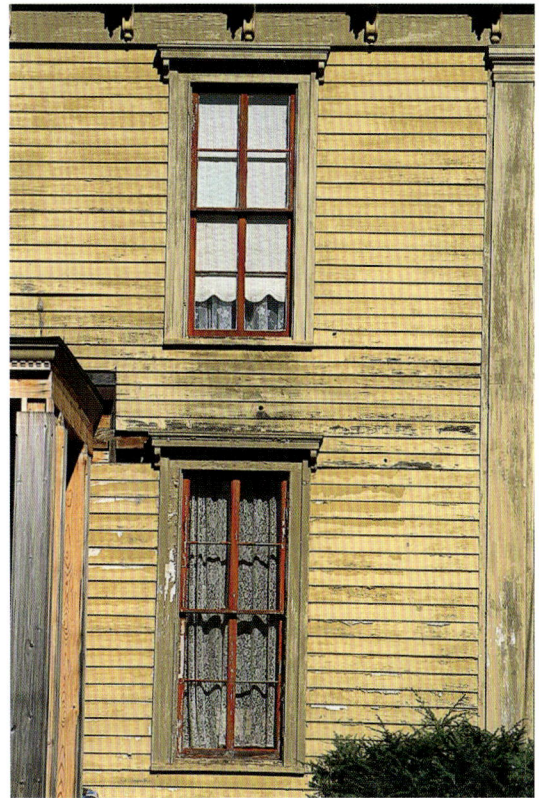

WI-47　有顶部装饰条的四格双悬窗，19世纪美国带有护墙板的房屋
Four-over-four double-hung windows with crown molding, 19th-c. American clapboard house

窗户种类

197

摄影：Austin, Patterson, Disston

WI-48　双悬窗：七格装饰窗；带有百叶窗的单格窗
Double-hung windows: seven-over-seven decorative lights; one-over-one with shutters

WI-49　从顶部到底部：山墙拱窗；六格一单格双悬窗；六格一单格双悬窗，传统板房
Top to bottom: arched gable window; arched six-over-one double-hung windows; six-over-one double-hung windows, traditional shingle house

WI-50　双悬窗，19世纪带有护墙板的教堂
Double-hung windows, 19th-c. clapboard church

窗户种类

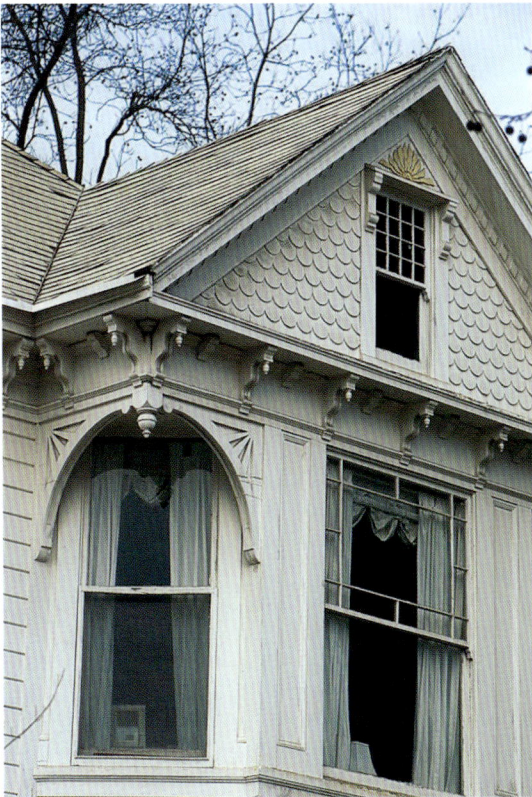

WI–51　双悬窗：上下窗均为单格（左）；十六格一单格窗
　　　　（山墙上）；上下均为单格装饰窗（右），19世纪带
　　　　有护墙板的房屋
　　　　Double-hung windows: one-over-one (left); sixteen-over-one
　　　　(gable); one-over-one decorative-lights (right), 19th-c. clapboard
　　　　house

WI–52　双悬窗：带有附壁竖框的二十格窗户（顶部）；上下
　　　　窗扇都是十二格的窗户，两侧是二十格一单格窗户，
　　　　20世纪木瓦房屋
　　　　Double-hung windows: twenty-over-twenty with pilaster
　　　　mullions (top); twenty-over-ones flanking twelve-over-twelve,
　　　　20th-c. shingle house

WI–53　有木框双四格双悬窗和防风窗，19世纪檐板房屋
　　　　Four-over-four double-hung windows with wood frames and
　　　　modern storm windows, 19th-c. clapboard house

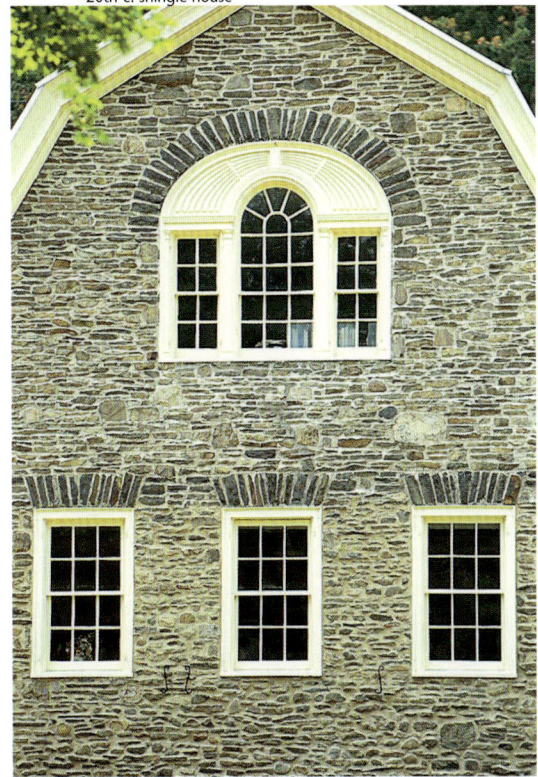

WI–54　顶部：固定的帕拉迪奥式窗；底部：双六格双悬窗，
　　　　美国石砌房屋
　　　　Top: fixed Palladian window; bottom: six-over-six double-hung
　　　　windows, American stone house

窗户种类

摄影：The Preservation Society of Newport County

WI-55　镶有曲面玻璃的弯曲的双悬窗；顶部：双十五格窗；底部：二十五格一单格窗，20世纪美国木瓦房
Curved double-hung windows with slumped glass. Top: fifteen-over-fifteen; bottom: twenty-five-over-one, 20th-c. American shingle house

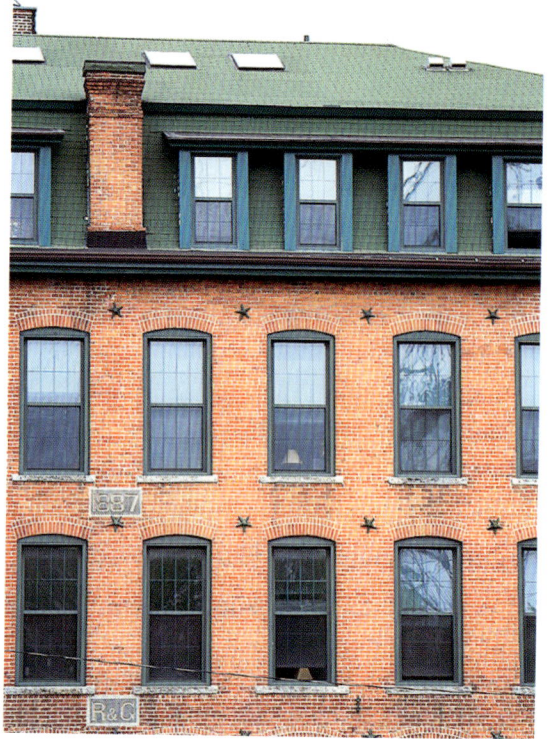

WI-56　带有内置窗格条的金属或者聚乙烯塑料材质的双悬窗，19世纪砖砌的工业建筑
Double-hung metal or vinyl replacement windows with internal muntins, 19th-c. brick industrial building

WI-57　单格双悬的金属或聚乙烯塑料替换窗，19世纪砖砌的公寓建筑
One-over-one double-hung metal or vinyl replacement windows, 19th-c. brick apartment building

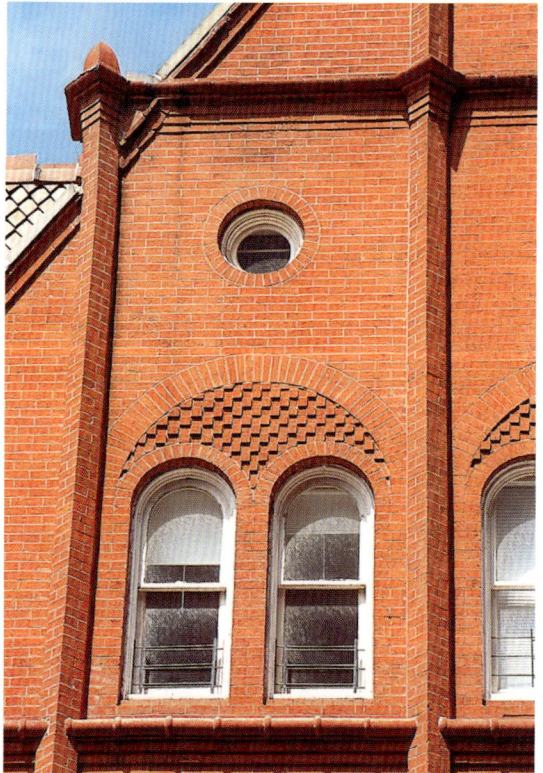

WI-58　带有固定窗的单格双悬窗，拱形的上部窗扇；顶部有固定的牛眼窗
One-over-one double-hung, arched top sashes with fixed lights; fixed bull's eye window above

窗户种类

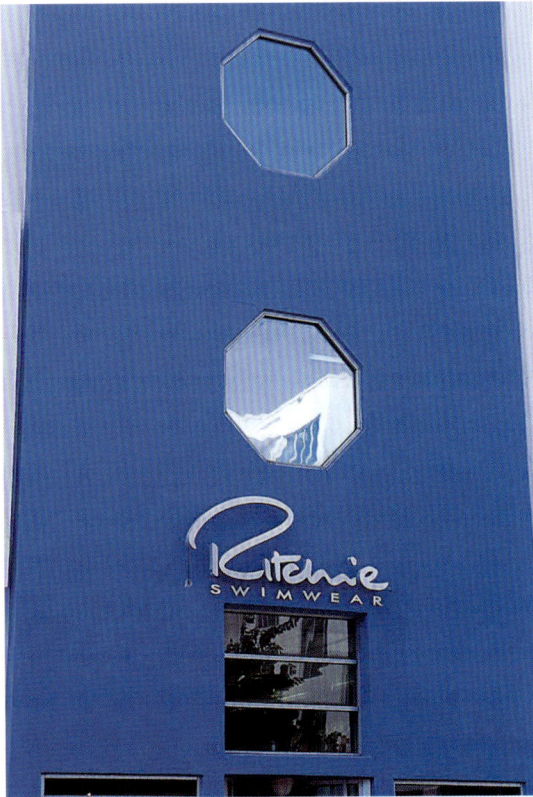

WI-59　固定的六边窗，现代商业建筑
Fixed hexagonal windows, modern commercial building

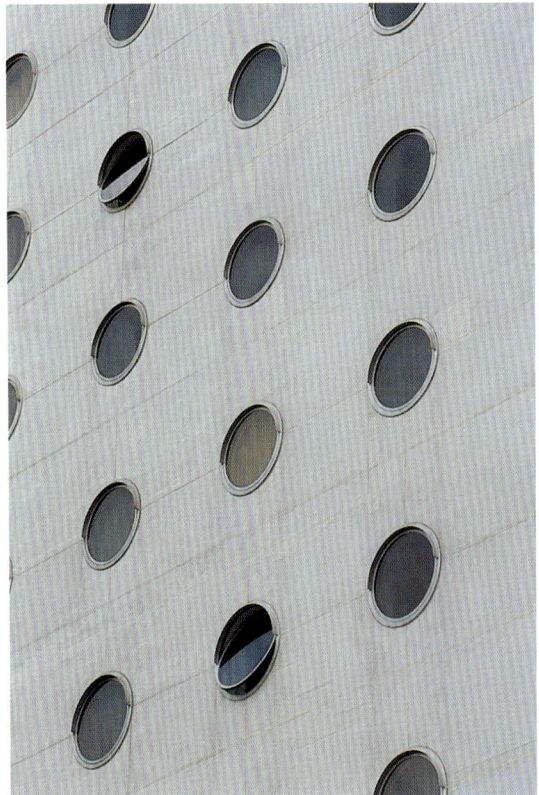

WI-60　圆形不锈钢外框的上悬窗，现代商业建筑
Circular awning windows with stainless steel frames, modern commercial building

WI-61　有叶形装饰以及花和彩带装饰大理石外框的椭圆形牛眼窗，19世纪的大楼
Oval bull's-eye windows with foliated marble frames and festoons, 19th-c. mansion

摄影：The Preservation Society of Newport County

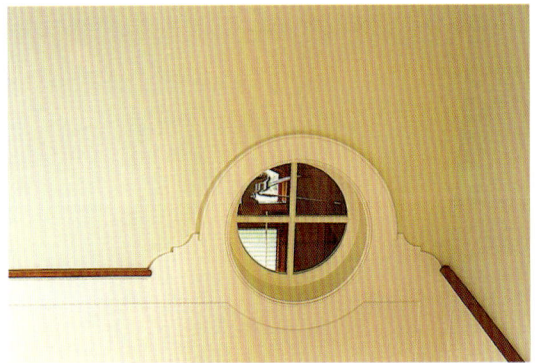

WI-62　被窗格条分开的圆形天窗，有一个扁平的木外框，现代房屋
Circular clerestory window divided by muntins, with a flat wood profiled frame, modern house

摄影：Austin, Patterson, Disston

WI-63　被窗格条分开的圆窗，带有雕刻石质外框
Circular window divided by muntins, with a carved stone frame

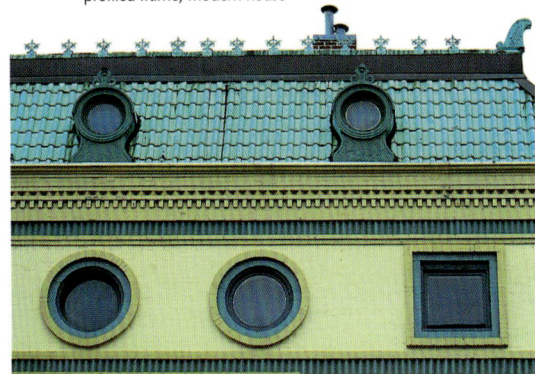

WI-64　牛眼窗式的屋顶窗和壁窗，还有一个方窗
Bull's-eye dormer- and wall windows, with one square window

WI-65 被散射式格窗条分隔的圆窗，石砌山墙顶部有一个通风口
Circular window divided by radiating muntins with ventilation opening above in stone gable wall

WI-66 被窗格分隔开的椭圆形窗户，白杉木瓦山墙上有扁平的木壳
Oval window divided by muntins, with flat wood casing in white cedar-shingle gable wall

WI-67 有石头花饰窗格的玫瑰花窗
Rose window with stone tracery

WI-68 由扁平窗格条分隔开的椭圆形窗户；叶形装饰石框上饰有两侧有花彩装饰的悬臂托梁
Oval window divided by flat muntins; foliated stone frame with ancon flanked by festoons

窗户种类

WI-69 上部气窗带有花格装饰的固定窗，现代房屋
Fixed windows with lattice in top lights, modern house

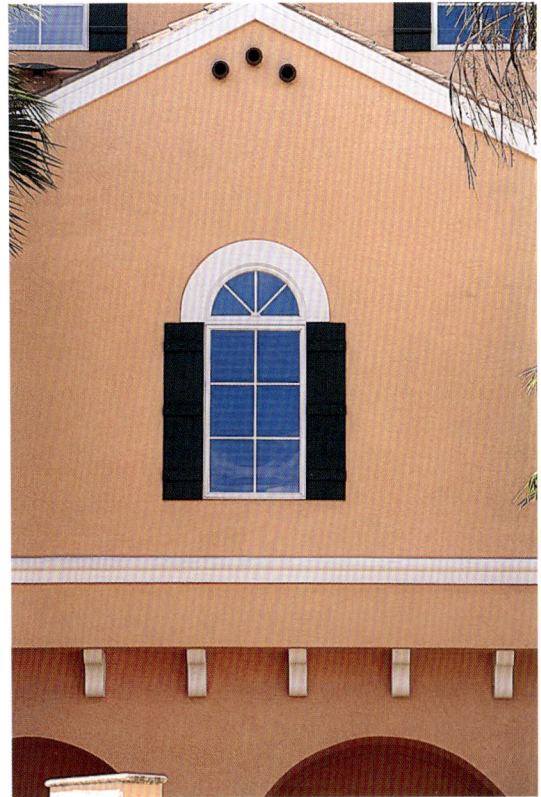

WI-70 拉毛水泥山形墙上带有内置窗格条、扇形窗和百叶窗的固定窗
Fixed window with internal muntins, fanlight, and shutters in stucco gable wall

WI-71 椭圆形、长方形和拱形的固定窗，地中海风格的现代房屋
Oval, rectangular, and arched fixed windows, Mediterranean-style modern house

WI-72 拱形和几何形排列的固定窗，现代房屋
Arched and geometric fixed windows, modern house

窗户种类

WI-73　几何形状的固定窗户，现代房屋
Geometric fixed windows, modern house

WI-74　弯曲墙面上的不锈钢框中的固定彩色窗户，现代商业建筑
Fixed tinted windows in stainless steel frames on curved wall, modern commercial building

WI-75　固定窗和边窗；有托架的遮阳篷，现代房屋的露台墙面
Fixed windows and sidelights; awnings with brackets, patio wall of modern house

窗户种类

摄影：Austin, Patterson, Disston

WI-76　石质凸窗，18世纪英国房屋
Stone oriel, 18th-c. English house

WI-77　多边形凸窗：六格、双悬、木框、砖基和外展的立接
缝屋顶，传统房屋
Cant-bay window: six-over-six, double-hung, with wood casings,
brick base, and flared standing-seam roof, traditional house

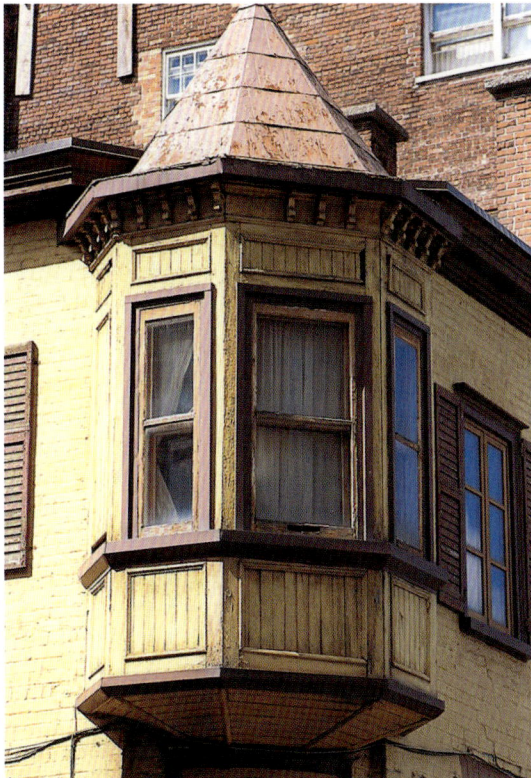

WI-78　墙角木凸窗：单格、双悬、木质窗扇和外框、圆锥形
的平接缝屋顶，19世纪的美国多功能建筑
Wood corner oriel: one-over-one, double-hung, wood sashes
and casings, conical flat-seamed copper roof, 19th-c. American
mixed-use building

WI-79　砖块双层角窗，下层为单格双悬窗，左侧为带有装饰
性气窗的拱窗，19世纪商业建筑
Brick corner two-story oriel: one-over-one double-hung windows
below and arched window with decorative lights (left), 19th-c.
commercial building

凸窗与角窗

WI-80 有固定窗的单层和双层凸窗，现代房屋
Single- and two-story oriels with fixed windows, modern house

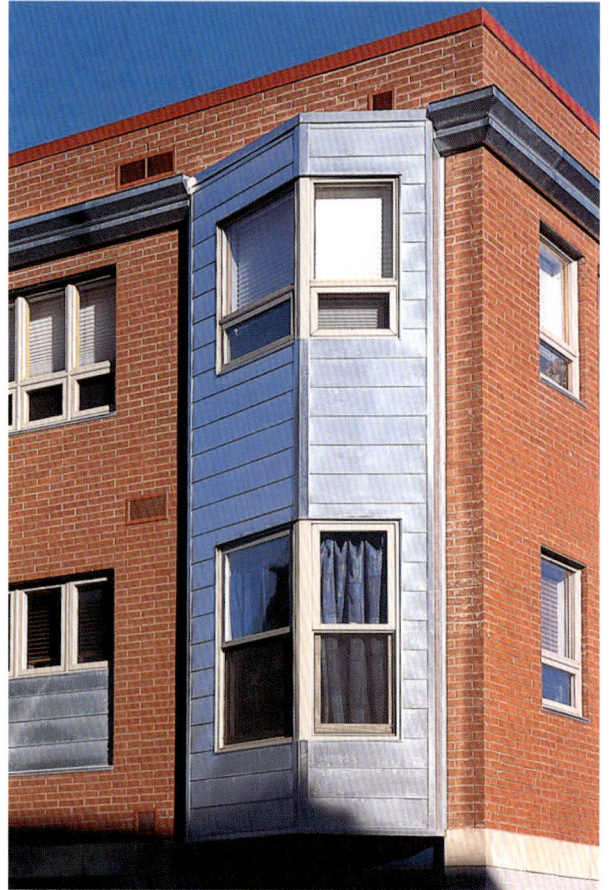

WI-81 金属覆层的双层角窗；顶部：有上悬气窗的固定窗；底部：
单格双悬窗扇，现代公寓房屋
Metal-clad two-story angled bay. Top: fixed windows with awning lights;
bottom: one-over-one double-hung sashes, modern apartment house

摄影：Roger Bartels Architects, Dobyan & Dobyan Custom Builders

WI-82 每边有三扇窗户的多边形凸窗，圆锥形木瓦顶，现代房屋
Cant-sided bay with three windows per side, conical shingle roof, modern house

凸窗与角窗

WI-83　双层多边形凸窗：浇铸或压制式金属装饰下的固定窗户，公寓房屋
Two-story cant-bay windows: fixed lights below ornamental cast or pressed metalwork, apartment house

WI-84　左：双开窗；右：多边形凸窗，固定的两侧窗和气窗，中间为平开窗
Left: double-casement window; right: cant-oriel with fixed side and transom lights and center casement window, 19th-c. mixed-use building

WI-85　石砌多边形凸窗：镶有花饰的窗格下为铅条镶嵌平开窗；上部有栏杆；窗户下有浮雕装饰带
Stone cant-oriel: leaded casement windows under tracery lights; roof balustrade; bas-relief ornamental band below window

凸窗与角窗

WI-86 角窗：带有金属窗框、金属竖棂和金属窗格条的固定窗户，现代商业建筑
Corner window: fixed lights with metal sashes, muntins, and mullions, modern commercial building

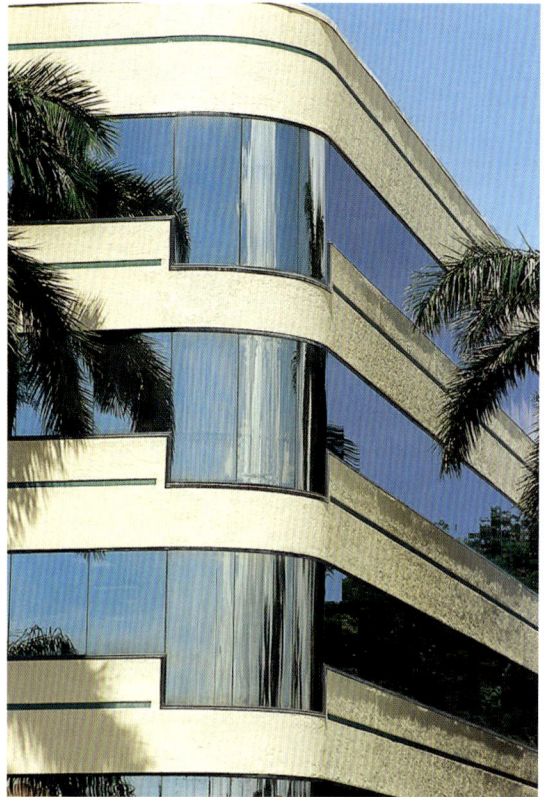

WI-87 有带状窗的弯曲角窗；彩色镜面玻璃，现代商业建筑
Curved corner windows with strip windows;tinted plate glass, modern commercial building

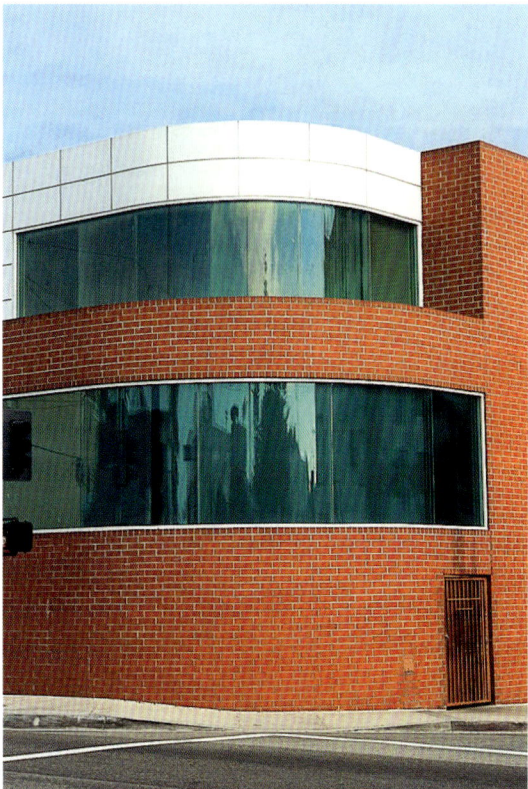

WI-88 弯曲的带状角窗，现代商业建筑
Curved corner strip windows, modern commercial building

WI-89 角窗：带有侧面窗格和气窗的双开窗，装饰艺术风格公寓
Corner windows: double casement with side lights and split transom lights, Art Deco apartment house

凸窗与角窗

WI-90　有固定窗的特殊的凸窗，现代商业建筑
Idiosyncratic bay with fixed lights, modern commercial building

WI-91　角窗：带有固定上扇窗的单悬窗，装饰艺术风格公寓
Corner windows: single-hung sashes with fixed upper lights, Art Deco apartment house

WI-92　带上悬窗和固定窗的凸窗，带有单坡顶和托臂，现代公寓
Oriel with awning or fixed windows, shed roof, and corbels, modern apartment house

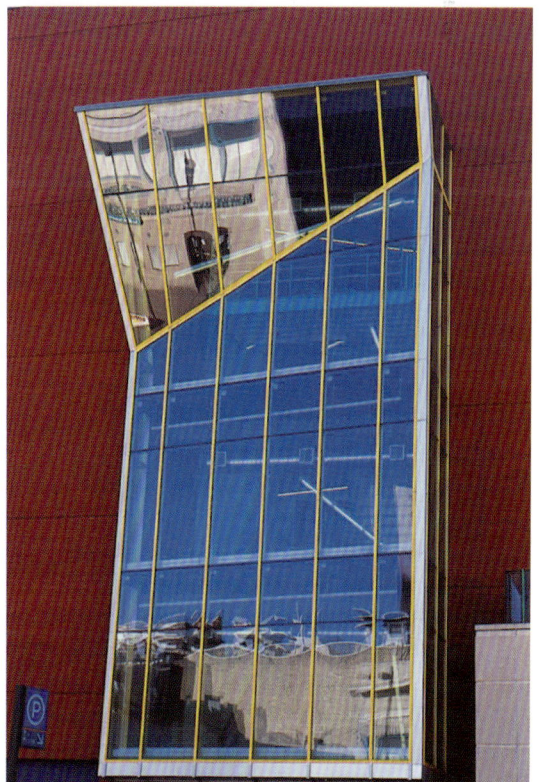

WI-93　上部为玻璃镜面的独特的固定凸窗，现代商业建筑
Idiosyncratic fixed-light bay with mirror-glass top section, modern commercial building

凸窗与角窗

摄影：Austin, Patterson, Disston

WI-94 弧形拱气窗和上部有木瓦"拱楔块"和"拱心石"的法式门，现代木瓦房
Segmental-arch transom light with shingle voussoirs and keystone above French door, modern shingle house

WI-95 附墙阳台下，带有造型的石质顶窗位于两侧有窗格的法式落地门上
Shaped stone transom light over French door with side lights under engaged balcony

WI-96 有固定铅条镶嵌窗格和石框的帕拉迪奥式窗中的扇形窗
Fanlight in Palladian window with fixed leaded lights and stone frame

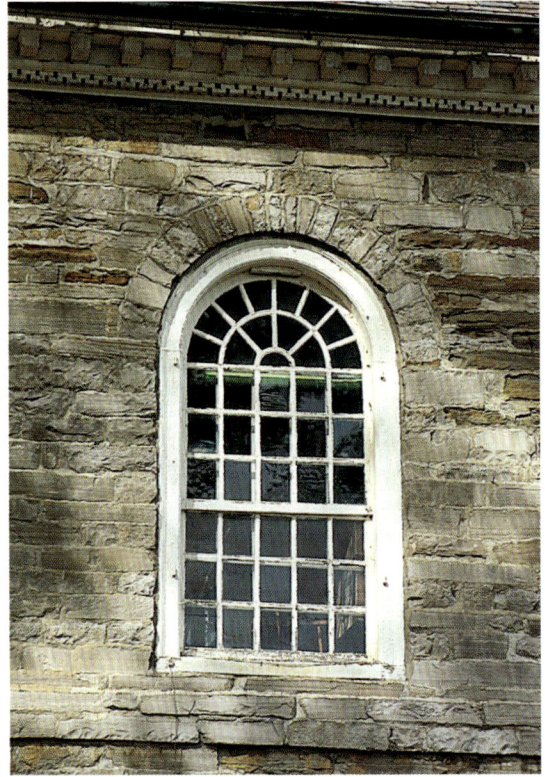

WI-97 十五格双悬木窗上的扇形窗
Fanlight over fifteen-over-fifteen double-hung wood window

顶窗与扇形窗

WI-98　带有装饰金属格栏的扇形窗
Fanlight with ornamental metal grille

WI-99　有拱心石和附壁柱的石框中的法式门上的半圆拱顶窗，英国房屋
Semicircular-arch transom lights over French doors in stone surround with keystones and pilasters, English house

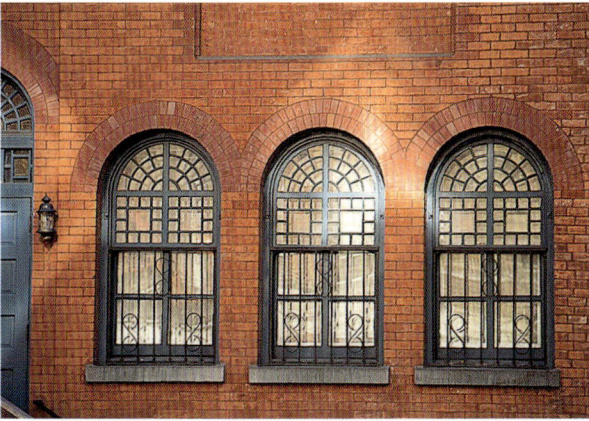

WI-100　方形固定窗格上的半圆拱扇形窗，有单格双悬窗框和窗门，19世纪联体别墅
Semicircular-arch fanlights over square fixed lights, with one-over-one double-hung sashes and window gates, 19th-c. townhouse

WI-101　平开窗上的铅条镶嵌矩形气窗；由托臂支撑着遮檐，工艺美术房屋
Leaded rectangular transom lights over casement windows; hood supported by brackets, Arts and Crafts house

WI-102　从顶部到底部：矩形气窗；固定窗；成对的单格双悬窗和固定的边窗，现代公寓
Top to bottom: rectangular transom lights; fixed lights; pairs of one-over-one double-hung windows and fixed side lights, modern apartment house

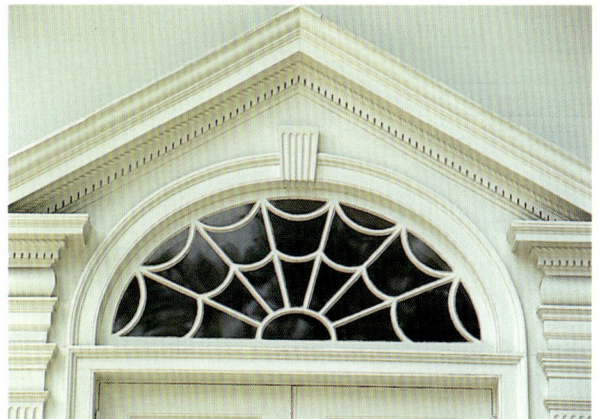

WI-103　三角墙下的蝙蝠翼扇形窗
Bat-wing fanlight under pediment

顶窗与扇形窗

WI-104　弓形三角墙顶部，带有以两侧的螺形支架作为装饰的双悬窗，19世纪美国房屋
Segmental pediment-capped window molding with consoles flanking double-hung windows, 19th-c. American house

WI-105　拉毛水泥砖墙中的双悬窗四周的扁木框，传统房屋
Flat wood casing around double-hung windows in stucco and brick wall, traditional house

摄影：Austin, Patterson, Disston

WI-106　双开窗四周的扁木框
Flat wood casing around double-casement windows

窗口装饰和窗户布局

WI-107 单格双悬窗四周有窗耳的扁木框；被托臂支撑着的窗檐，美国木板房
Flat wood casing with crossettes around one-over-one double-hung windows; hood supported by brackets, American clapboard house

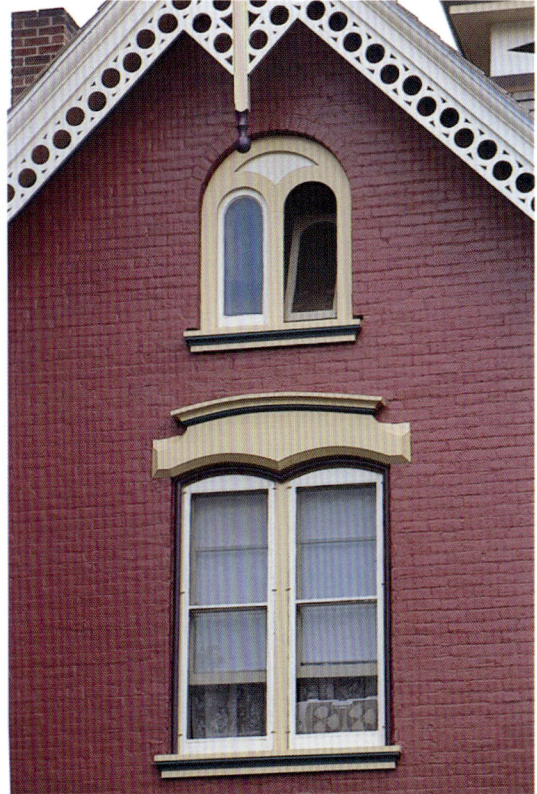

WI-108 扁平的装饰条与窗户的拱形轮廓相呼应，美国砖房
Flat trim echoing arch profiles of window openings, American brick house

WI-109 扁平的装饰条与窗户的拱形轮廓相呼应，19世纪美国多功能建筑
Flat trim echoing arch profiles of window openings, 19th-c. American mixed-use building

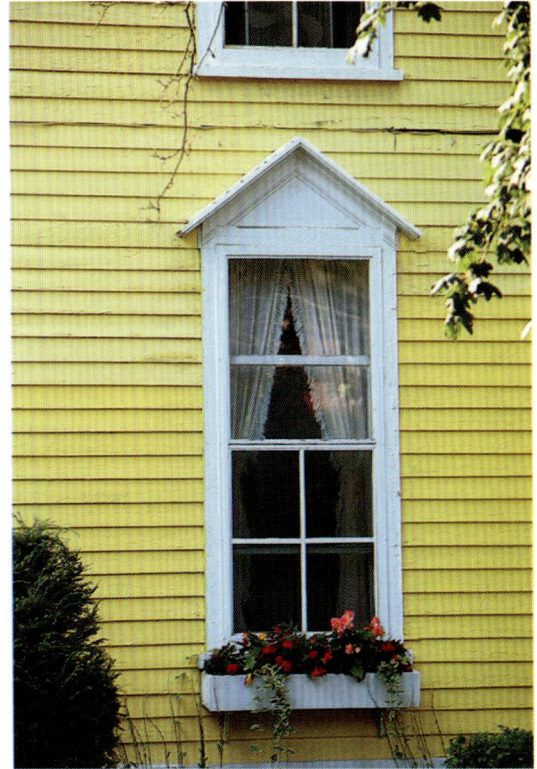

WI-110 双悬窗上的木质山墙形窗檐，美国木板房
Wood gable-shaped hood over double-hung window, American clapboard house

窗口装饰和窗户布局

摄影：Austin, Patterson, Disston

WI-111 不同的窗户布局：半木结构山形墙中的六格双悬窗，传统房屋
Variety of fenestrations: six-over-six double-hung windows in half-timber-style gable wall, traditional house

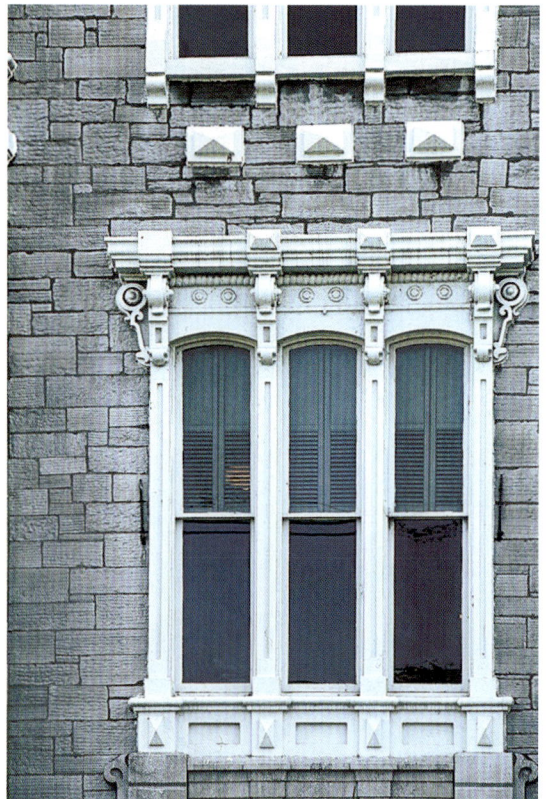

WI-112 装饰和线脚环绕着三扇双悬窗组；螺形支架支撑着上楣，19世纪美国石砌商业建筑
Ornament and molding around group of three double-hung windows; consoles supporting cornice, 19th-c. American stone commercial building

WI-113 带有图案的扁窗框包裹着双开窗
Profiled flat casing around double-casement window

摄影：Roger Bartels Architects, Dobyan & Dobyan Builders

WI-114 扁窗框包裹着角窗和水平窗，现代木瓦房
Flat casing around corner windows and horizontal window, modern shingle house

窗口装饰和窗户布局

WI-115　被相连的支柱夹着的山墙双挂老虎窗上方的圈形山墙
Round pediment over gable double-hung dormer window flanked by pairs of attached columns

WI-116　多边双悬凸窗和壁窗四周的装饰、线脚和栏杆；19世纪美国墙板房
Ornament, molding, and balustrade around double-hung windows in cant-bay and wall windows; 19th-c. American clapboard house

WI-117　三层山墙双挂窗的设置；扁木框，有皇冠形线脚的顶盖
Set of three gable wall double-hung windows; flat wood casings, with crown molding caps

窗口装饰和窗户布局

WI-118 规准砖平拱、楔形砖和"拱心石"的颜色与周围形成对比
Gauged-brick jack arch, brick quoins, and keystone in contrasting color

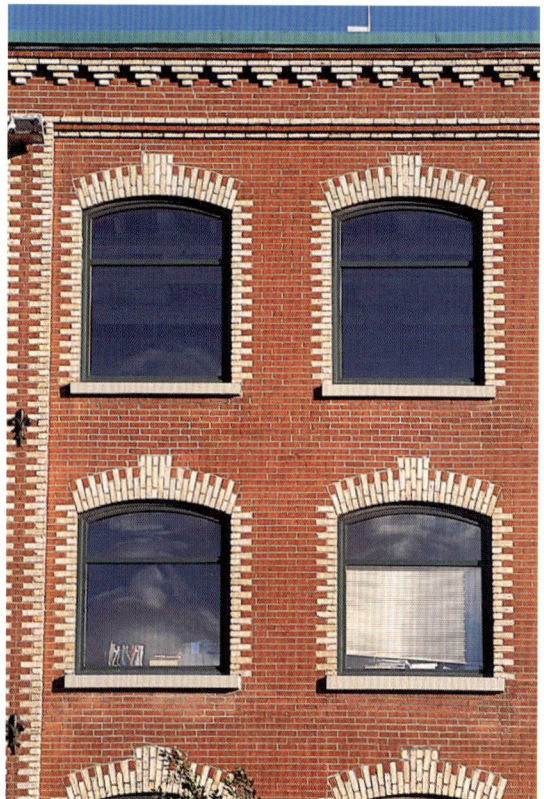

WI-119 带有楔形砖的弧形拱窗和"拱心石"的颜色与周围形成对比
Gauged-brick segmental-arch windows with brick quoins and keystones in contrasting color

WI-120 砖瓦带顺着弧形拱的轮廓
Masonry and tile band following profile of segmental arch

WI-121 层列碎石墙中的造型窗口，19世纪美国西南部大厦
Shaped embrasure in coursed-rubble wall,19th-c. American Southwestern mission

窗口装饰和窗户布局

WI-122 用砖拱和附壁柱作框的弧形拱平开窗
Segmental-arch-shaped casement window framed by brick arch and pilasters

WI-123 层列碎石墙面中的带有扁平拱心石的拱和楔形石形成双悬窗的外框
Stone flat keystone arches and quoins framing double-hung windows in coursed-rubble wall

WI-124 墙面石头窗楣、砖带、雕刻石框顶和双悬窗四周的砖带
Stone lintels and carved casing tops with stone and brickwork band around double-hung windows in brick wall

WI-125 上楣有皇冠线脚的古典风格的石窗框罩着下层的窗户；螺形支架支撑着窗台
Stone classical-style window casings with crown molding capping lower window; consoles supporting sill and cornice

摄影：The Preservation Society of Newport County

窗口装饰和窗户布局

WI-126　螺旋形石雕图案环绕着带有线脚式窗框的上层窗，可替换的窗户，20世纪早期公寓住宅
Stone cartouches flanking upper-story windows with molded casings, replacement windows, early 20th-c. apartment house

WI-127　风景窗四周的石头装饰和被壁柱隔开的边墙；石窗框嵌入葱形拱之中
Stone ornament around picture window and side windows separated by pilasters; top stone casing pierced with ogee arches

WI-128　四周有花环的陶制上楣环绕着陶制框窗；螺旋形装饰下的封闭窗
Terracotta cornice with garlands around windows with terracotta casings; blind window under cartouche

窗口装饰和窗户布局

WI-129　方形粗石乱砌石墙的平拱窗
Windows with jack arches in roughly squared random stone wall

WI-130　有浮雕装饰和装饰性拱心石的石砌或预浇铸混凝土窗框，20世纪早期公寓
Stone or cast concrete casings with relief ornament and ornamented keystones, early 20th-c. apartment house

WI-131　有线脚式石框窗的三扇凹窗和单线脚窗舌的设置，20世纪早期商业建筑
Set of three recessed windows with stone molded casings and single molded sill, early 20th-c. commercial building

窗口装饰和窗户布局

WI-132 古典石窗框包着百叶窗，老式罗马建筑
Stone classical casing around shuttered window, old Roman building

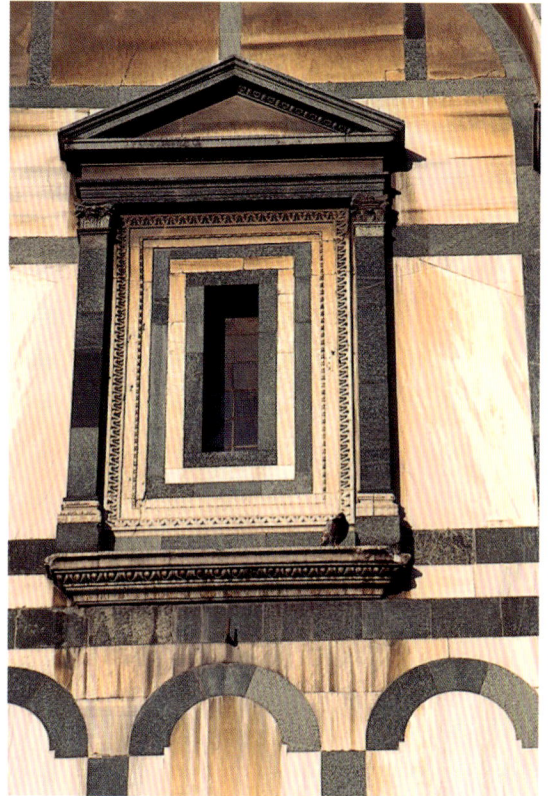

WI-133 大理石山墙和壁柱框住了有彩色大理石带的雕刻大理石窗框，12世纪意大利教堂
Marble pediment and pilasters framing carved marble casing with polychrome marble bands, 12th-c. Italian church

WI-134 石头折衷主义风格装饰框住的金属替换窗，公寓
Stone eclectic ornament flanking metal replacement window, apartment house

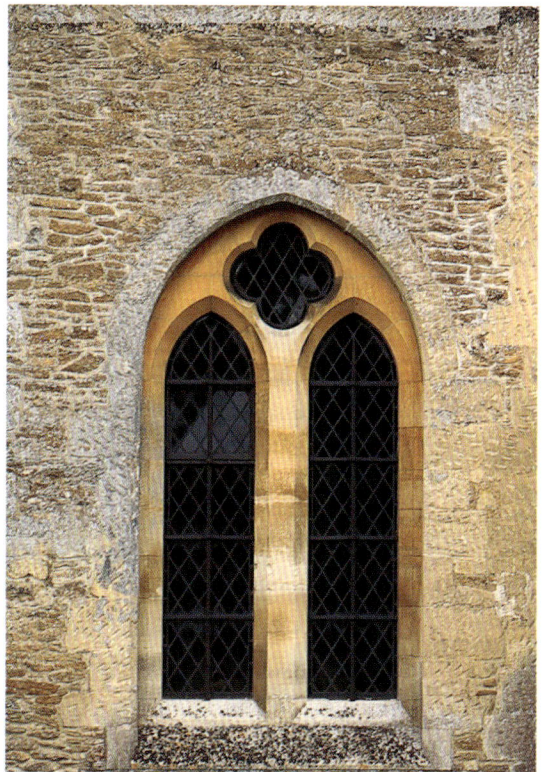

WI-135 石墙上的尖拱内是一对柳叶格子窗和四瓣窗，带有石材边框，15世纪英国修道院
Stone pointed segmental arch framing pair of leaded lancet windows with stone casings and leaded quatrefoil, 15th-c. English abbey

窗口装饰和窗户布局

WI-136 石框外的彩色装饰陶制外框，公寓建筑
Polychrome ornamented architectural terracotta frame around stone casings, apartment building

WI-137 带有被附壁柱围住的螺旋形装饰的奇特的新哥特式上下层窗间墙面，公寓建筑
Fanciful neo-Gothic spandrel with cartouche flanked by corbelled pilasters, apartment building

WI-138 带有附墙栏杆的半木风格山墙上的壁柱围住的带石框的三扇窗设计
Set of three dormer windows with stone casing flanked by pilasters in half-timber-style gable with engaged balustrade

WI-139 有螺旋装饰的卷轴饰三角墙，附壁柱和带边窗的窗下栏杆，公寓
Scrolled pediment with cartouche, engaged columns, and balustrade below window with side lights, apartment house

窗口装饰和窗户布局

WI-140 装饰性赤褐色砂石拱石环绕着椭圆形拱窗
Ornamented brownstone voussoirs around semicircular-arch window

WI-141 上下层窗间墙面上的新哥特式装饰加盖窗板，20世纪早期商业建筑
Neo-Gothic ornament in spandrels and capping window panels, early 20th-c. commercial building

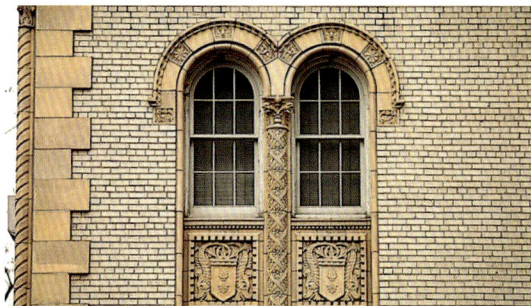

WI-142 装饰性半圆拱加在被单附壁柱分开的双拱窗上，底部有螺旋装饰
Ornamental semicircular arches capping pair of arched windows separated by single engaged column; bottom cartouche panels

WI-143 连拱、附壁柱和彩色陶制或石头拱腹环绕着带有石窗框的窗户，19世纪法国混合功能建筑
Series of arches, pilasters, and polychrome terracotta or stone spandrels framing windows in stone casing, 19th-c. French mixed-use building

WI-144 有框的浅浮雕门楣中心和窗户四周装饰性的壁柱；带有上釉瓷砖的封闭拱窗
Framed bas-relief tympanums and ornamented pilasters around windows; blind-arch window with glazed tile

WI-145 不同种类的线脚和装饰环绕着异形窗上方的方形平开窗
Assortment of moldings and ornament around square casement windows over profiled fixed window

窗口装饰和窗户布局

WI-146　有浮雕设计和金属边框的上下层窗间墙
Metal spandrel with relief design and metal casings

WI-147　浇铸浮雕设计和金属边框上下层窗间墙面
Metal spandrels with cast relief designs and metal casings

WI-148　固定尖头窗和圆形窗四周的石拱
Stone arch around fixed lancet-shaped and circular lights

WI-149　嵌壁式石框中，上方为铜嵌板的带有金属窗框格的窗户，20世纪商业建筑
Copper panels capping windows with metal sashes in recessed stone casings, 20th-c. commercial building

窗口装饰和窗户布局

WI-150　带阳台的威尼斯风格窗户四周的浮雕装饰嵌板和附
壁柱，20世纪公寓房屋
Relief ornamental panel and pilasters around variation of Venetian-style window with balcony; 20th-c. apartment house

WI-151　水平堆叠的木板环绕着固定窗，现代商业建筑
Horizontal stacked wood board surround around fixed-light window, modern commercial building

WI-152　嵌入拱形上方的拉毛水泥墙中的方形平开窗，西班牙
殖民风格彩色图案
Square casement window recessed in stucco wall above arch;
painted Spanish-colonial–style motifs

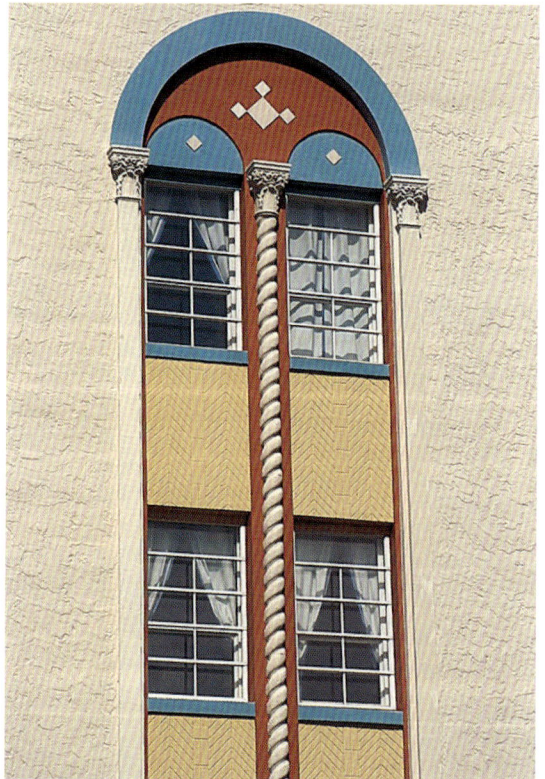

WI-153　双层威尼斯风格窗户；喷漆拉毛水泥和浇铸混凝土
装饰，装饰艺术风格酒店
Two-story variation of Venetian-style window; painted stucco
and cast-concrete ornament, Art Deco hotel

窗口装饰和窗户布局

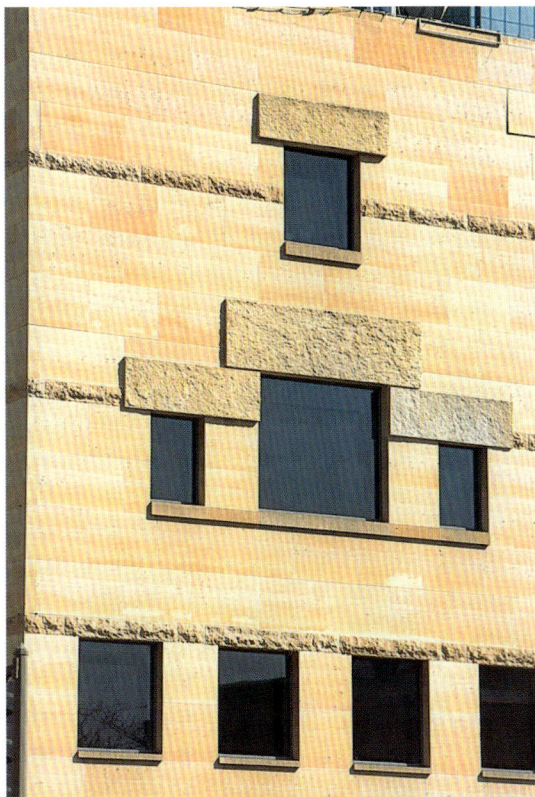

WI-154　不同的窗口布局，固定窗，现代商业建筑
Variety of fenestrations of rectangular, probably fixed, windows, modern commercial building

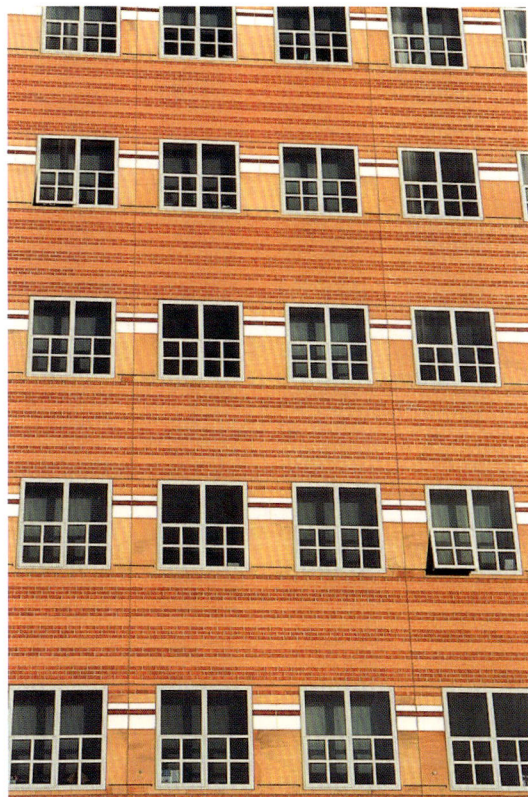

WI-155　带状外立面上由带有金属框的窗户构成的重复的窗口布局设计
Repetitive fenestration of metal-framed windows in banded facade design, modern commercial building

WI-156　带边窗的风景窗的重复的窗口布局设计和带有楔形砖的双悬窗（右），现代商业建筑
Repetitive fenestration of picture windows with side lights and double-hung windows (right) with brick quoins, modern commercial buildin

WI-157　上下层窗间墙面上连接着上下窗的金属设计
Metal design in spandrels connecting upper and lower windows

窗口装饰和窗户布局

WI-158　上下层窗间墙面有凸出嵌板的三道窗格条的重复的窗口布局设计，现代商业建筑

Repetitive fenestration of three-mullion windows with raised panel spandrels, modern commercial building

WI-159　带状窗口布局设计：拱窗列和方窗列；金属替换窗，19世纪工业建筑

Banded fenestration: row of arched windows and rows of square windows; metal replacement windows, 19th-c. industrial building

WI-160　由上下层窗间石墙面分开的嵌壁式方形金属框固定窗，现代商业建筑

Stone spandrels dividing recessed square metal fixed windows, modern commercial building

WI-161　石头边框中带有双扇百叶窗的固定窗组，现代商业建筑

Group of fixed windows with two louvered ventilation panels in stone surround, modern commercial building

WI-162　有上下层窗间金属墙面和边框的方窗链接设计，现代商业建筑

Interlocking design of square windows with metal spandrels and surround, modern commercial building

WI-163　有开放式和封闭式遮光板的重复的窗口设计，现代公寓建筑

Repetitive fenestration with open and closed window shades, modern apartment building

窗口装饰和窗户布局

WI-164 镜面玻璃条形窗，现代商业建筑
Mirror-glass strip windows, modern commercial building

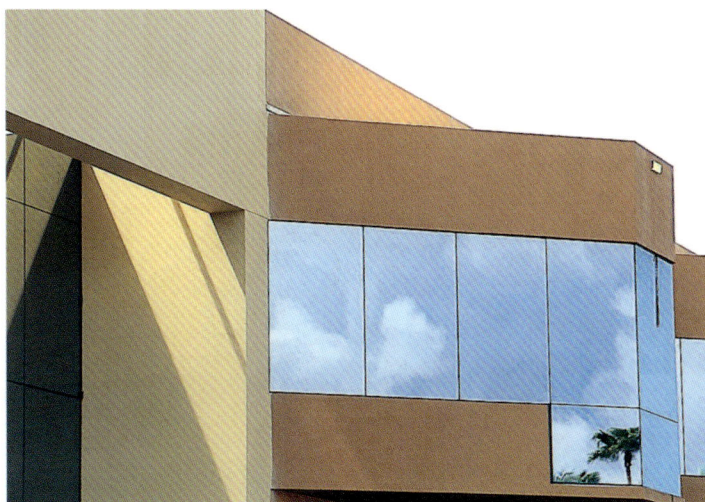

WI-165 镜面玻璃条形窗，现代商业建筑
Mirror-glass strip windows, modern commercial building

WI-166 被实体嵌板分开的带状窗户，现代公寓建筑
Ribbon windows separated by solid panels, modern apartment building

窗口装饰和窗户布局

WI-167　不锈钢窗护栅
Stainless steel window gate

WI-168　有装饰花边的喷漆铸铁窗栏
Painted wrought-iron window grille with decorative border

WI-169　四周为蔓藤图案设计的镂空菱形花纹装饰黄铜窗护栏
Brass window grille with arabesque design around pierced diaper-work

WI-170　有菱形图案的铸铁窗护栅，现代公寓房屋
Wrought-iron window gates with diaper pattern, modern apartment house

窗护栅与栅栏

WI-171　华丽的金属护栏
Patinated filigree metal grille

WI-172　六边形和六角星装饰互连图案的黄铜护栏
Brass grille with interlocking hexagon-and-star pattern

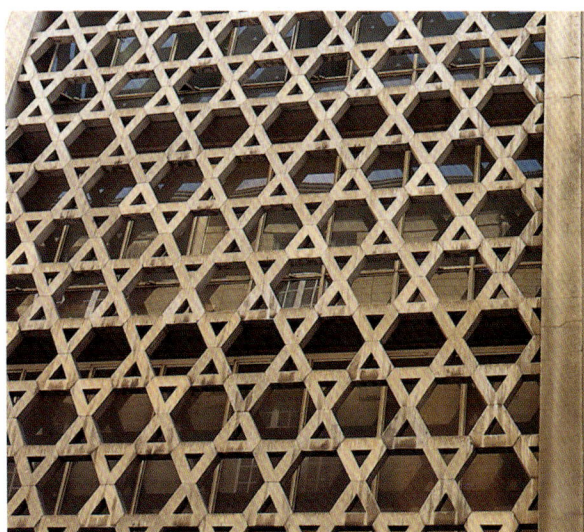

WI-173　有菱形图案的镂空大理石护栏
Pierced marble grille with diaper pattern

WI-174　对角线设计的竹护栏
Bamboo grille in diagonal design

窗护栅与栅栏

WI-175 西班牙殖民风格的铸铁护栏
Wrought-iron window gate in Spanish-colonial style

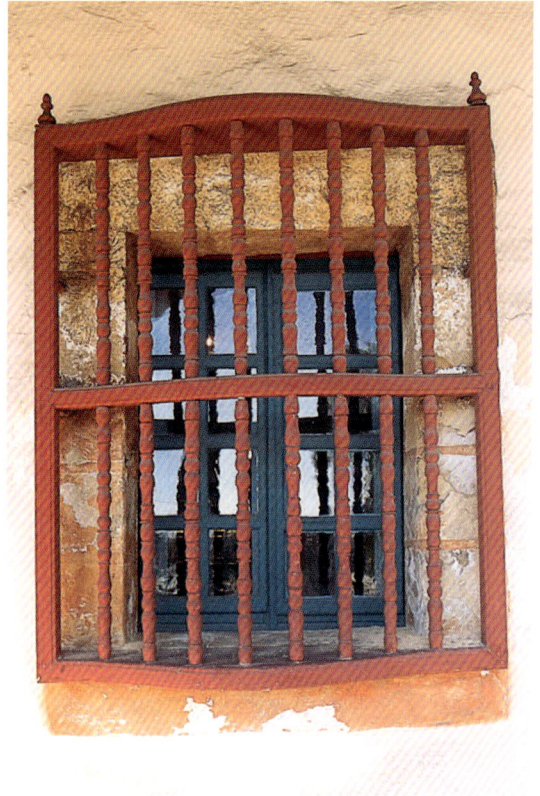

WI-176 车削的木制护栏，美国西南部建筑
Grille of turned wood bars, American Southwestern mission

WI-177 粗木护栏
Rustic wood grille

WI-178 传统日式格子窗栏
Traditional Japanese lattice window grille

窗护栅与栅栏

WI-179　金属百叶窗，19世纪工业建筑
Metal shutters, 19th-c. industrial building

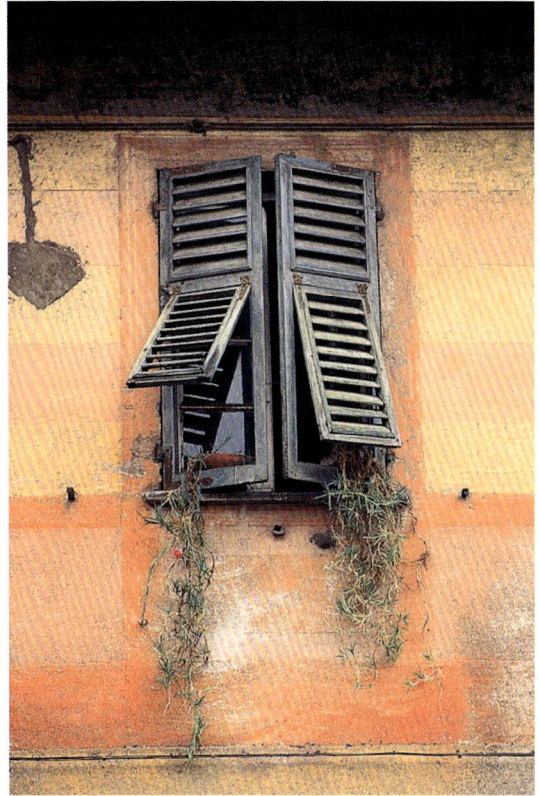

WI-180　被风化双悬百叶活动窗板
Weathered double-hinged louvered shutters

WI-181　风化的金属遮阳篷
Weathered metal awning

WI-182　成对的有外框的百叶活动窗
Pairs of framed louvered shutters

百叶窗

WI-183　有固定的上层百叶的百叶窗
Louvered shutters with fixed top shutters

WI-184　有固定的上层百叶的百叶窗
Louvered shutters with fixed top shutters

WI-185　双悬活动百叶窗
Double-hinged louvered shutters

WI-186　活动百叶窗
Louvered shutters

WI-187　粗面活动百叶窗
Rustic louvered shutters

WI-188　现代塑料活动百叶窗
Modern plastic louvered shutters

百叶窗

WI-189　粗面板条活动百叶窗板
Rustic battened shutters

WI-190　板条活动百叶窗板
Battened shutters

WI-191　有镂空设计的水平板百叶窗板
Horizontal-board shutters with pierced design

WI-192　有"Z"字形板条的活动窗板
Shutters with Z-battens

WI-193　带有板条的活动窗板和门
Battened shutters and door

百叶窗

WI-194 风化的喷漆金属活动窗板
Weathered painted metal shutter

WI-195 带有挖空图案的突面窗板
Raised-panel shutter with cutout

WI-196 粗面仓库活动窗板
Rustic warehouse shutters

WI-197 风化的喷漆木嵌板活动窗板
Weathered painted wood paneled shutters

WI-198　百叶式的镶嵌式活动窗板
Louvered and paneled shutters

WI-199　板材式活动窗板，日本房屋
Planked shutter, Japanese house

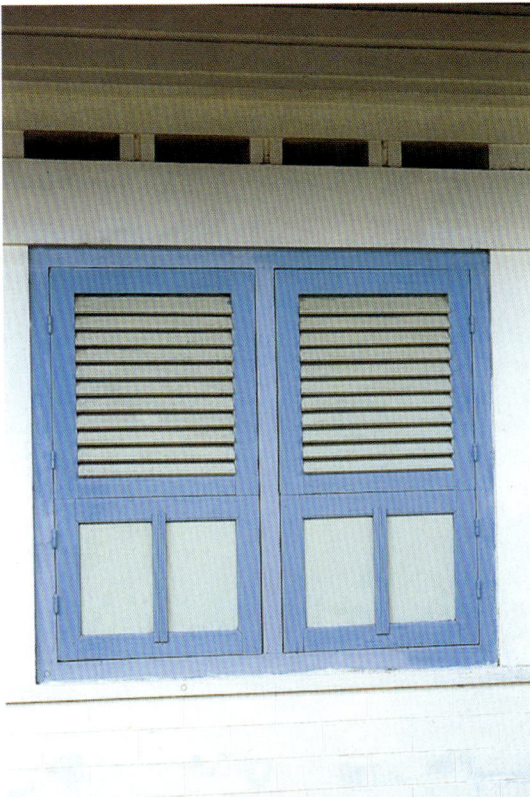

WI-200　有上部气窗口的百叶窗
Louvered shutters with upper lights

WI-201　滑动式格子窗，日本房屋
Sliding lattice shutter, Japanese house

百叶窗

WI-202　半透明遮阳嵌板，现代多功能建筑
Sun screens of translucent panels, modern mixed-use building

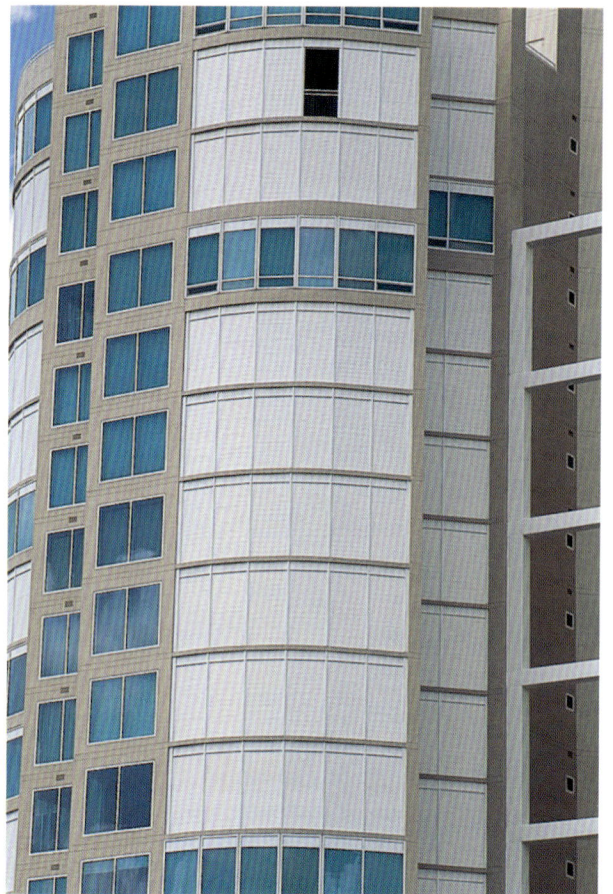

WI-203　防飓风百叶窗，现代公寓
Hurricane shutters, modern apartment house

WI-204　百叶天窗，现代商业建筑
Skylight shutters, modern commercial building

百叶窗

门廊

DW-1 3块木质门芯板，带有镶铅条的透视窗，黄铜把手，装饰金属板，门框、镶板和门的横木上镶有饰钉；弓形拱门
Three narrow panels, wood, with leaded lights, brass handle, decorative plates and studs on stiles, panels, and rails; segmental-arch doorway

DW-2 21块木质门芯板，带有镶铅条的透视窗，带有邮件插槽，在门框和门的横木上装饰有饰钉
Twenty-one panels, wood, with leaded light, mail slot, and decorative studs on rails and stiles

DW-3 6块木质门芯板，双扇门板上雕刻有菱形图案，装饰性把手，底部门板饰有铆钉，并带有格子气窗
Six panels, wood; double doors with carved diamond pattern in panels, ornament framing handles, studded bottom panel, and transom grille

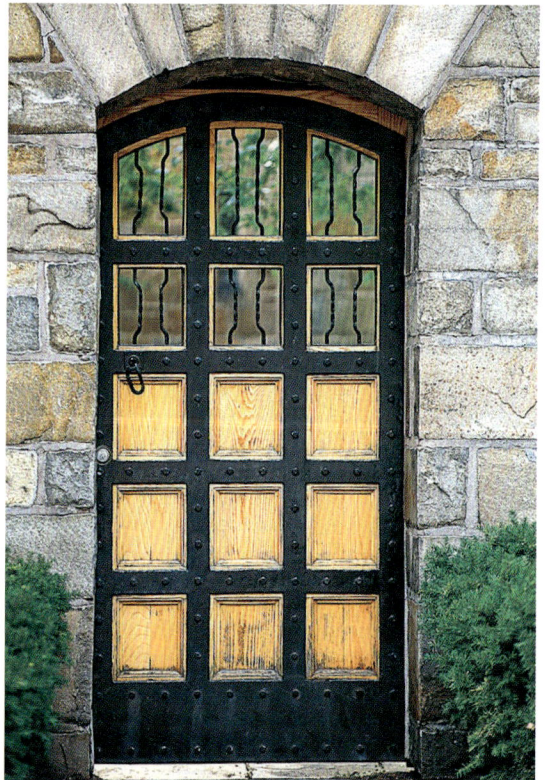

DW-4 15块木质和开放式格子镶板，喷漆门框和横木上饰有铆钉；入口上方为弓形拱
Fifteen wood and open-grilled panels, studs on painted stiles and rails; segmental-arch doorway

镶板门

DW-5 4块镶板，带有突出线脚和突面镶板的木质双开门
Four panels, wood double doors with flush molding and raised panels

DW-6 饰有几何图案木质镶板；左：蚀刻玻璃；右：镶嵌木
Wood with geometric patterned panels. Left: etched glass; right: inlaid wood

DW-7 6块镶板，嵌有厚木板和铸铁把手的未抛光的双扇木门；金属门框带有横楣窗和边窗的
Six panels, unfinished wood double doors with planked panels and wrought-iron handles; metal door-frame with transom light and side lights

DW-8 双镶板，饰有浅浮雕装饰的带有立体感的木质双扇门；门的上方为饰有漩涡卷饰拱心石的弧面石拱
Two panels, double doors with carved bas-relief ornament and trompe l'oeil wood finish. Cambered stone arch doorway with cartouche keystone

镶板门

DW-9 镶有边饰的双镶板；楣石上有花状平纹的附壁柱门框；镶玻璃气窗和边窗

Two panels with applied trim; pilaster doorframe with anthemion on lintel; transom light and side lights

DW-10 四块有对角线装饰的厚木镶板，带有锻铁把手的喷漆门，带有护墙板的房屋

Four diagonally planked panels, painted door with wrought-iron handle, clapboard house

DW-11 单门芯板的双扇车库门和单门芯板门，未抛光的木贴面板

Single-panel double garage doors and single-panel door; unfinished wood veneer

镶板门

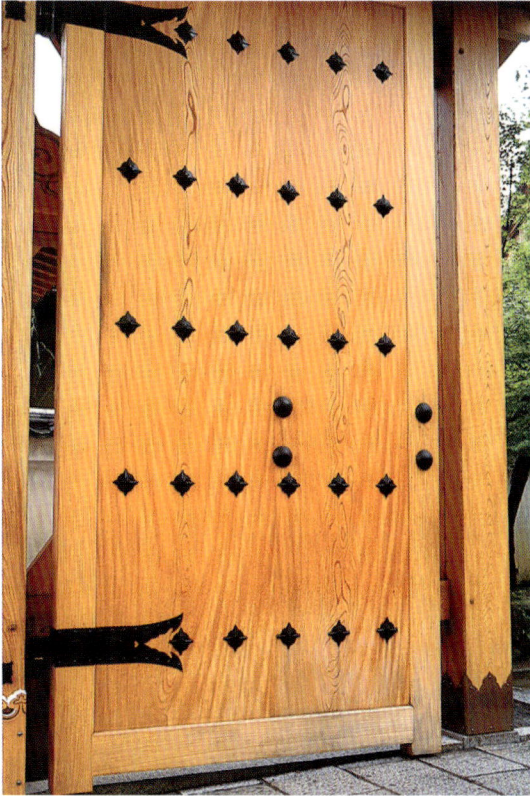

DW-12 单嵌板，有锻铁把手的实木，日本庭院门
Single-panel, book-matched wood with wrought-iron hardware, Japanese courtyard gate

DW-13 有双交叉支撑的风化双镶板木门，门的中心横木上有黄铜饰件和球形门把手，日本庭院门
Weathered two-panel wood with double cross-braces, brass hardware, and doorknob in center stile, Japanese courtyard gate

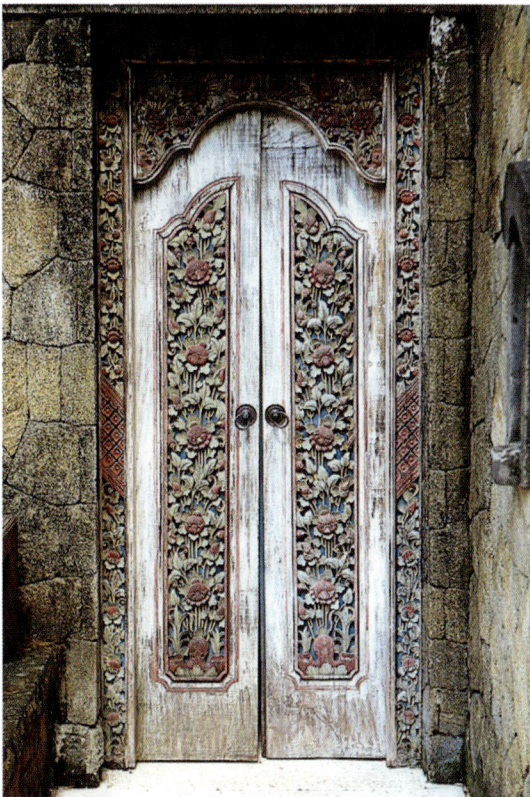

DW-14 有彩色浅浮雕图案的单镶板双扇门，印度尼西亚庭院门
Single-panel double doors with polychrome bas-relief panels, Indonesian courtyard gate

DW-15 5块镶板，有菱形窗栅的木质双扇门，日本寺庙
Five panels, wood double doors with diamond grille, Japanese temple

镶板门

摄影：Austin, Patterson, Disston

DW-16 带有突面镶板的6块镶板木门；边窗，住宅前门
Six-panel mahogany door with raised panels; side lights, residential front entry

DW-17 带有突面镶板的6块镶板木门，雕刻线脚和半圆气窗
Six-panel wood door with raised panels, carved molding, and half-round transom light

DW-18 带有突面镶板的车库木门
Wood garage doors with raised panels

镶板门

DW-19 有突出脚线的双镶板木质双扇门；上部镶有玻璃，下部是分割的嵌板；中心板上的投信槽门上方有气窗

Two-panel wood double doors with flush molding; upper panels glass, lower panels divided; letter slot in center stile; split transom light

DW-20 喷漆木门；有九块玻璃的螺纹玻璃嵌板；齐平的木嵌板；半露的门框

Painted wood door; ribbed-glass panel with nine lights; wood panels flush-molded; pilaster frame

DW-21 上层为玻璃镶板，下层为齐平木镶板的双扇木门，弧形拱的入口上方中有分开的气窗

Wood double doors with glass upper panels and flush-molded lower panels; split transom light in segmental-arch doorway

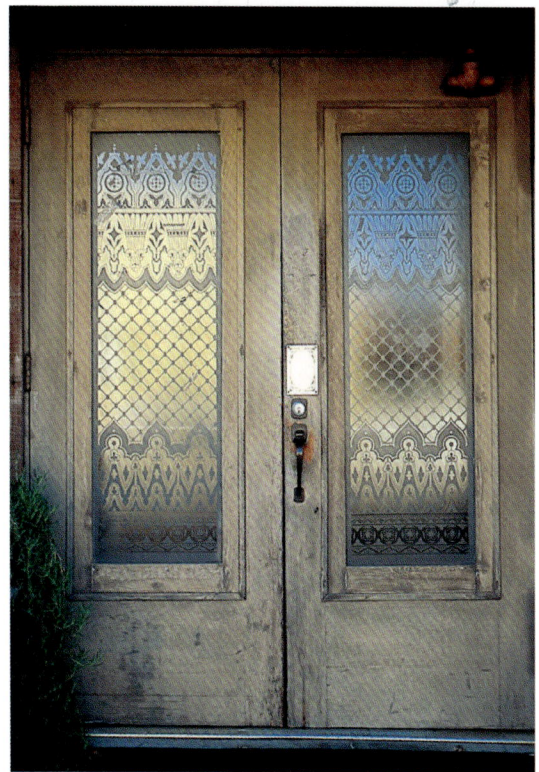

DW-22 风化的双扇木门；单块玻璃镶板上装饰有霜状和蚀刻的Z字形和花纹的图案

Weathered wood double doors; single glass panels frosted and etched with zigzag and diaper patterns

镶板门

DW-23 有中心壁柱的拱形双扇木门；拉毛水泥外围，美国西南部地区风格

Arch-shaped wood double doors with center pilaster; stucco surround, American Southwestern mission style

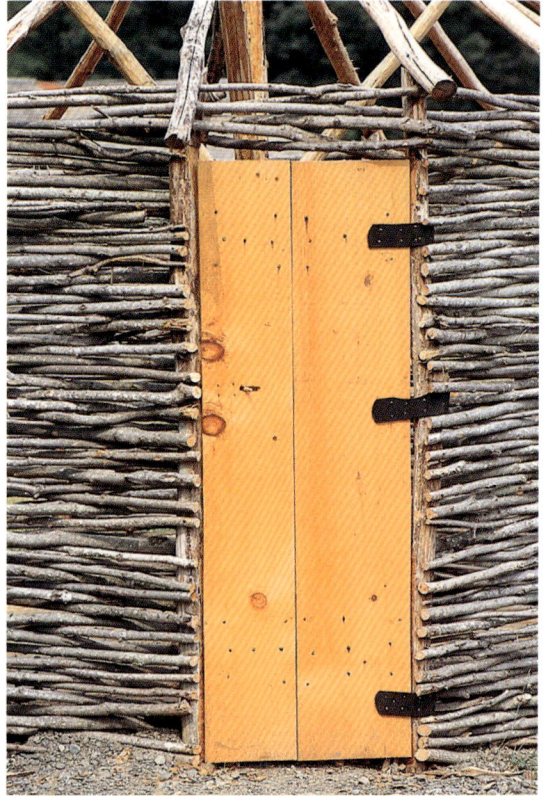

DW-24 有锻铁颌的平锯厚木板；编织的木墙，18世纪谷仓

Plain-sawn plank door with wrought-iron strap hinges; woven wood walls, 18th-c. corn crib

DW-25 有双Z形撑臂的弦切厚木板，农场棚屋

Plain-sawn planks with double Z-brace, farm shed

DW-26 粗木门楣下的风化厚木门；谷仓和阁楼的粗面和拉毛水泥山墙

Weathered plank doors under rough-hewn timber lintel; rustic stone and stucco gable wall of barn and hayloft

板条门

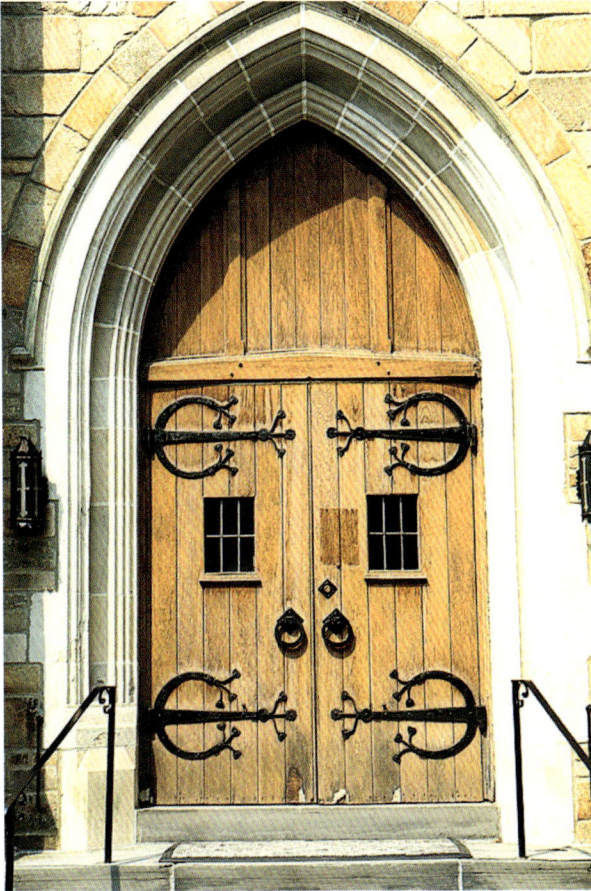

DW-27 有格子窗和装饰性锻铁把手的橡木门；拱形镶板
Oak door with leaded lights and ornate wrought-iron strap-hinges; arched transom panel

DW-28 有格子窗和装饰性黄铜长页铰链的橡木门；弓形拱
Oak door with leaded-glass light, ornate brass strap hinges; segmental arch

DW-29 有水平支撑的着色庭院木门，安装在喷涂拉毛水泥墙上
Stained wood courtyard gate with horizontal braces, set in painted stucco wall

DW-30 风化谷仓上的喷漆木挡板和悬挂式双扇门
Painted wood siding and hanging double doors on weathered barn

DW-31 粗石墙面上带有窗户和弯曲的Z字形撑柱的车库双扇木门，艺术手工房屋
Wood double garage doors with lights and curved Z-braces in fieldstone wall, Arts and Crafts house

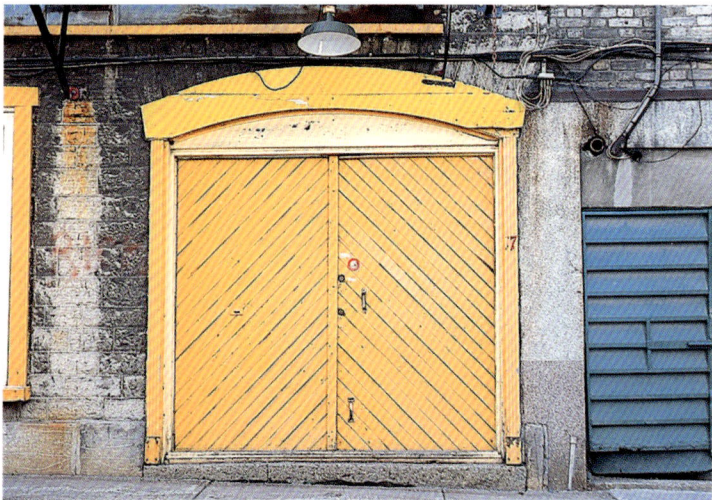

DW-32 斜板组成的车库双扇木门，位于弓形拱中
Wood double garage doors with diagonal planking, in segmental arch

DW-33 有粗制长页铁铰链和门闩的风化喷漆木门，农家院
Weathered painted wood-plank door with rustic iron strap hinges
and latch, farm courtyard

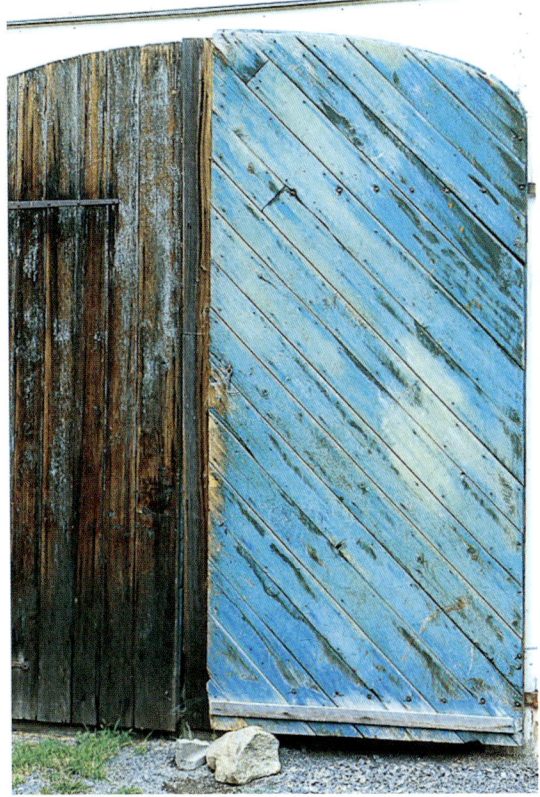

DW-34 有粗锯水平厚木板和喷漆斜厚木板的风化双扇木门
Weathered wood double doors with rough-sawn vertical planks
and painted diagonal planks

DW-35 风化的粉刷双扇木板门，混凝土砖车库
Weathered whitewashed wood-plank double doors, concrete-
block shed

DW-36 有单板支撑的粗面木板门，锻铁配件和双嵌板气窗；
平拱；16世纪或17世纪英国房屋
Rustic wood-plank door with single brace, wrought-iron hardware,
and two-paneled transom window; cambered arch; 16th- or
17th-c. England

DW-37 立体感强烈的错视画法木纹双镶板双扇门，带有立体的齐平线脚；弧形拱

Trompe l'oeil wood-grained two-paneled double doors, dimensional flush molding; segmental arch

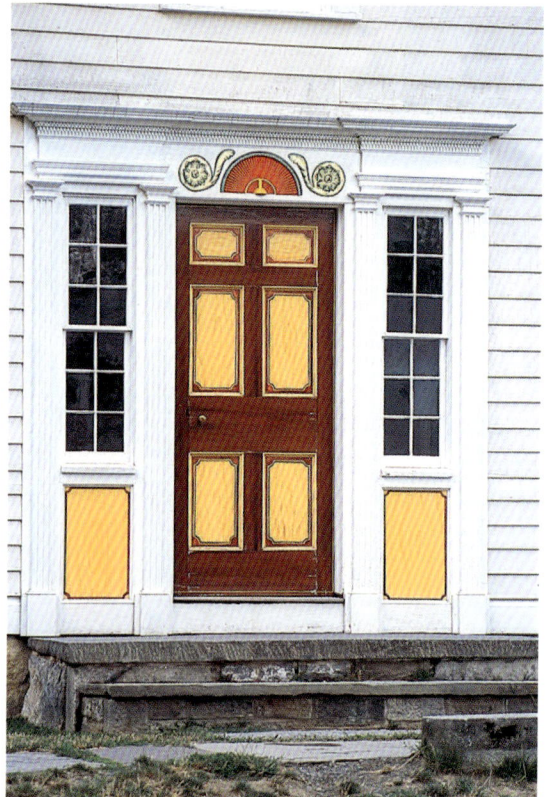

DW-38 有齐平线脚的六镶板美国殖民式风格门，门楣上有喷漆装饰和双悬边窗

Six-panel American Colonial-style door with flush molding, painted ornament on transom, and double-hung side lights

DW-39 有喷漆镶板和喷漆雕刻装饰的三镶板双扇门；弧形拱门楣，巴厘岛

Three-panel double doors with painted panels and painted carved ornament; segmental arch transom, Bali

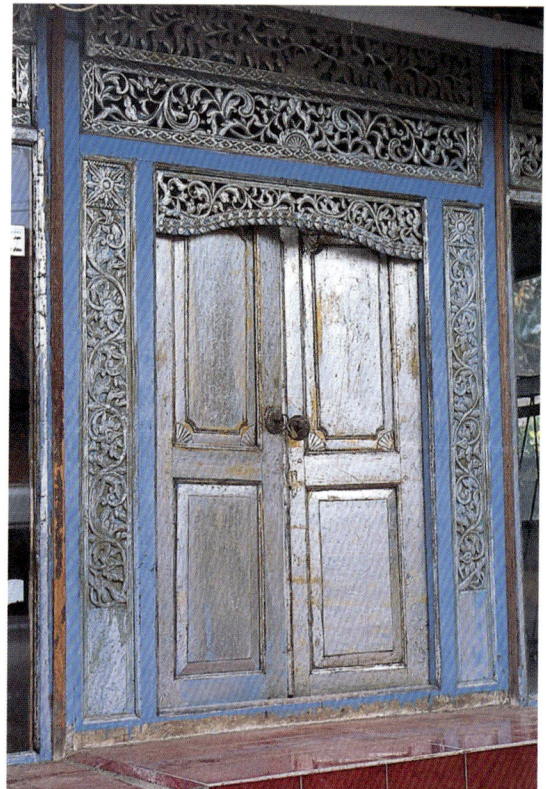

DW-40 装有突面镶板、雕花边板和镂空雕刻门楣的双镶板双扇门，全部喷有金属漆

Two-panel double doors with raised panels, carved side panels, and pierced carved transom, all with metallic paint

喷漆门

DW-41 顶部嵌有五角窗的喷漆镶板门，带有黄铜投信口，门在带有平拱的砖石入口中

Painted paneled door with pentagonal light in top panel, brass letter slot, in brick entry with jack arch

DW-42 有喷漆镶板和纺锤形栅格的门；粉刷的泥砖入口带有外露的木质门楣，美国西南部地区建筑

Door with painted panels and grille of turned spindles; stucco adobe doorway with exposed timber lintel, American

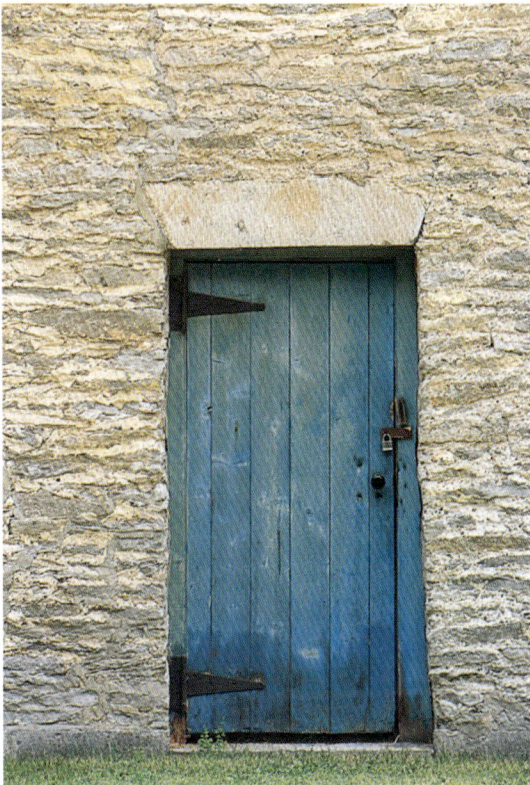

DW-43 带铁合页的风化喷漆板条门

Weathered painted wood batten door with iron strap hinges

DW-44 有喷漆门芯板的风化的双扇门，入口嵌在带有裸露木质门楣的拉毛水泥墙中

Weathered double doors with painted panels; doorway recessed in stucco reveal with exposed timber lintel

DW-45 黄铜三镶板双扇门，带有玻璃的镶板和带有线脚的外框；黄铜百叶门顶窗嵌板
Brass three-paneled double doors, panels with lights and framed with molding; louvered brass transom panel

DW-46 全玻璃的不锈钢双扇门，玻璃上带有圆形窗格条
Stainless steel double doors with full glass in circular muntins or grille

DW-47 全玻璃铝板旋转门
Aluminum-clad revolving door with full glass

DW-48 锻铁青铜嵌板平开门和旋转门；镂空图案的门顶装饰栅
Cast-bronze paneled hinged and revolving doors; pierced transom grille

DW-49 黄铜双扇门；有扇形窗格条的全玻璃设计；黄铜门框，装饰
艺术风格商业建筑
Brass double doors; full glass with radial muntins or grille; brass doorframe;
Art Deco commercial building

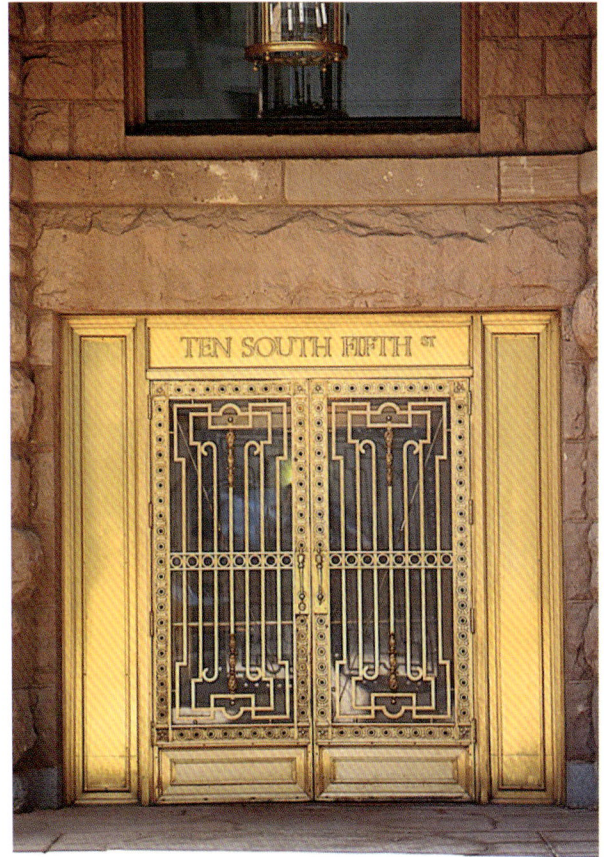

DW-50 黄铜双扇门；装饰性黄铜格栏后为全玻璃设计；黄铜嵌板在
门框两侧
Brass double doors; full glass behind ornate brass grille; brass panels
flanking doorframe

DW-51 玻璃砖墙中的无框玻璃双扇门
Frameless glass double doors set in glass-block wall

金属门

DW-52 带"填褥"嵌板的金属镀层门；门开口上方的砖块被竖砌砖带所环绕
Metal-clad door with quilted panels; brick surround with rowlock course above door opening

DW-53 有涂鸦的镀锌金属安全卷门
Rolling galvanized-metal security door with graffiti

DW-54 有条纹玻璃顶窗的风化车库钢门，现代房屋
Weathering-steel garage doors with textured glass lights, modern house

金属门

DW-55 焊接的喷漆金属抽象拼贴画应用到门和门框之中，门带有肌理的金属门槛

Welded and painted metal collage applied to door and frame, with textured metal threshold

DW-56 带有不锈钢饰件的喷漆金属门；刷饰面不锈钢门框镶有顶窗

Painted metal door with stainless steel hardware; brushed stainless steel doorframe with transom light

DW-57 大门墙面内的铁门；装饰性金属饰品中有带有金属包层的侧门基座；16世纪日本城堡

Iron door inset in gate wall; bases of adjacent reveals clad in decorative metal work; 16th-c. Japanese castle

金属门

DW-58 嵌壁式入口中有三扇顶窗的不锈钢框全玻璃双扇门；石头外围和台阶；现代商业建筑

Full glass double doors in stainless steel frame with three-light transom in recessed entryway; stonesurround and steps, modern commercial building

DW-59 带有顶窗的金属框中的双窗刷饰面铝门，现代公寓房屋

Brushed aluminum door with two lights in metal frame with transom light, modern apartment house

DW-60 带有旋涡饰和椰树装饰的亮漆铝质纱门

Enamel-painted aluminum screen door with scrollwork and palm ornament

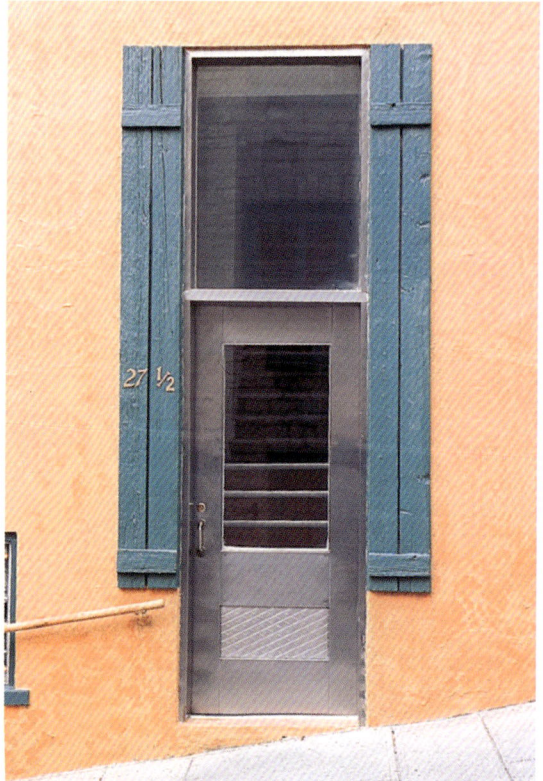

DW-61 带有单块玻璃板和顶窗的不锈钢门；门框两侧为挡板，旧市区建筑的现代入口

Stainless steel door with single light and transom light; frame flanked by shutters, modern doorway in older urban building

DW-62 椭圆形顶窗安在有单块玻璃板的木镶板门上；门框两侧有陶立克式附壁柱和古典式柱头，19世纪公寓

Oval transom light over wood paneled doors with single lights; doorframe flanked by Doric pilasters and entablature, 19th-c. apartment house

DW-63 六窗格双顶窗在有中心接线柱的九窗格喷漆双扇木门上方，传统房屋

Six-light double transom panels over nine-light painted wood double doors with center post, traditional house

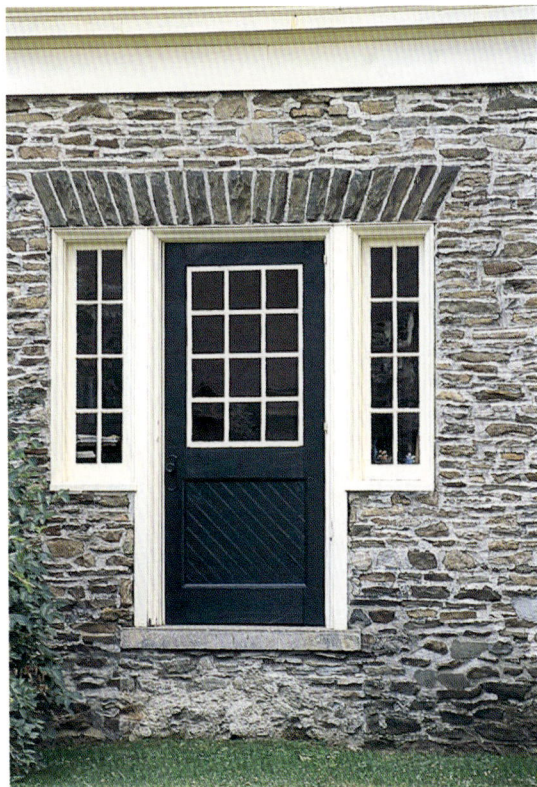

DW-64 边窗夹着的十二窗格喷漆木质镶板门，散石墙面中的楔形砖砌平拱；美国殖民风格乡村建筑

Side lights flanking twelve-light painted wood paneled door; gauged-brick jack arch in fieldstone wall; American colonial village building

DW-65 半圆形入口处的双扇木门和边窗；有拱心石和拱端托的石拱

Wood double doors and side lights in semicircular doorway; stone arch with keystone and imposts

摄影：Austin, Patterson, Disston

顶窗与边窗

DW-66 木质喷漆入口带有三窗格边窗、八窗格顶窗和单玻璃板镶板门，商业建筑
Painted wood doorway with eight-light transom panel over three-light side lights and single-light paneled door, commercial building

DW-67 全玻璃双扇门和边窗，上方带有三窗格顶窗，现代房屋
Full glass double doors and side lights with three-light transom, modern house

DW-68 全玻璃双扇门和双边窗，现代房屋
Full glass wood double doors and double side lights, modern house

顶窗与边窗

DW-69 有八边形玻璃和镶铅条玻璃顶窗的双扇镶板门；投信槽；凹槽壁柱；冠顶饰

Paneled double doors with grilles over octagonal lights with leaded glass transom in cambered arch; letter slot; fluted pilasters; crown molding

DW-70 带尖头玻璃板和十二窗格顶窗的双扇镶板门和边窗；投信槽；爱奥尼亚式壁柱和柱头

Paneled double doors and side lights with lancet-shaped lights and twelve-light transom; letter slot; Ionic-style pilasters and entablature

DW-71 单玻璃板门和边窗在带有弯曲木质门楣的原木中

Single-light door and side lights in rustic wood with curved timber eyebrow lintel

顶窗与边窗

DW-72 附墙圆柱上有科林斯式柱冠的装饰性拱形入口，扇形窗，拱门饰，螺旋形拱心石和卷轴三角墙

Ornate arched entry with Corinthian capitals on engaged columns, fanlight, archivolt, cartouche keystone, and scrolled pediment

DW-73 有台阶的古典风格大理石入口，科林斯式支柱和古典式柱头，以及有砖砌托饰的三角墙

Classical-style marble entry with steps, Corinthian columns and entablature, and pediment with block modillion

DW-74 有陶立克式双附墙圆柱支撑着齿状的新古典式入口，八镶板双扇门，蝙蝠翼扇形窗

Neoclassical entry with pairs of Ionic engaged columns supporting dentil entablature, eight-paneled double doors, bat-wing fanlight

DW-75 带有雕刻双扇木门的石质半圆拱；拱心石和托臂支撑着一个阳台，巴黎风格的公寓

Stone segmental arch with carved wood double doors; keystone and corbels supporting a balcony, Parisian apartment house

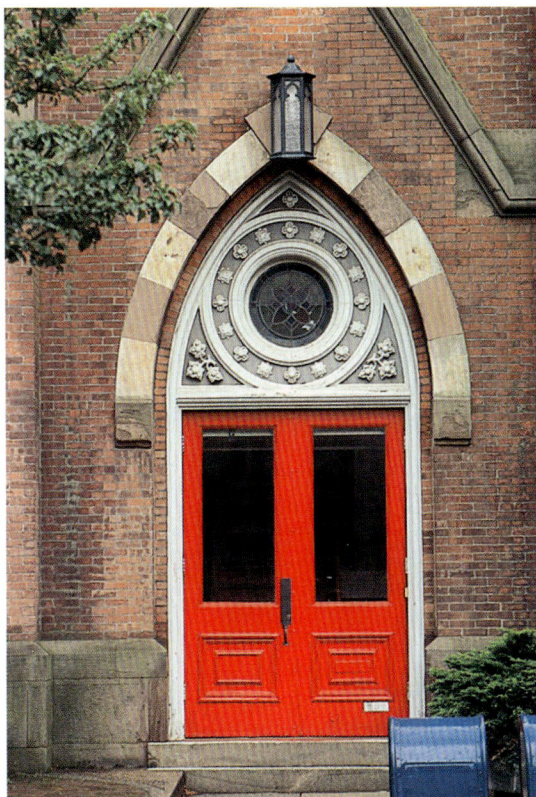

DW-76 带有牛眼窗的石雕门楣上方带状石中间的尖拱；双镶板单玻璃板双扇门，砖石结构教堂

Pointed arch in banded stone over carved stone transom panel with bull's eye window; two-paneled single-light double doors, brick church

DW-77 喷漆板门上有拱座的弧形拱；拉毛水泥面上有骷髅装饰，19世纪美国西南部地区

Segmental arch with molded imposts over painted plank door; skull ornament in stucco surround, 19th-c. American Southwestern mission

摄影：Lockwood-Mathews Mansion-Museum

DW-78 维多利亚风格的温室入口：喷漆四玻璃板木门和蚀刻顶窗；木壁柱和皇冠线脚；石头基座

Victorian conservatory entry: painted wood four-light door and transom light with etched glass; wood pilasters and crown molding; stone base

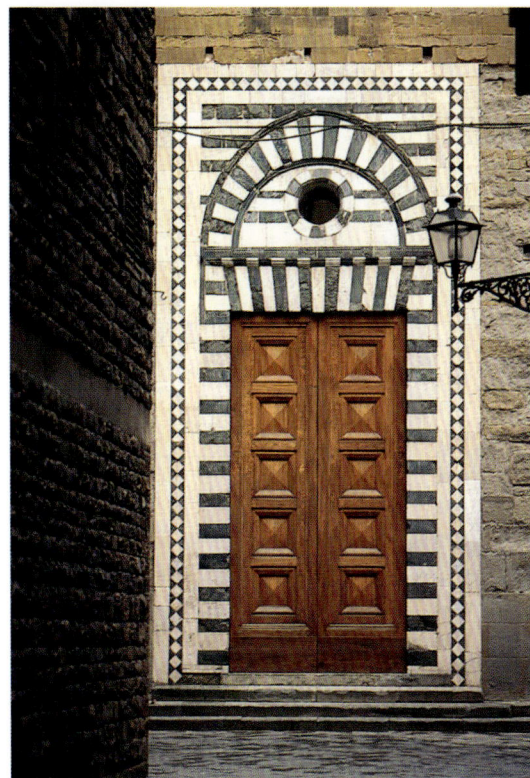

DW-79 有菱形饰带的大理石入口，佛罗伦萨拱，楔形砖砌平拱在雕刻双扇木门上方，16世纪意大利教堂

Marble entry with diamond band, Florentine arch, and gauged-block jack arch over carved wood double doors, 16th-c. Italian church

门廊

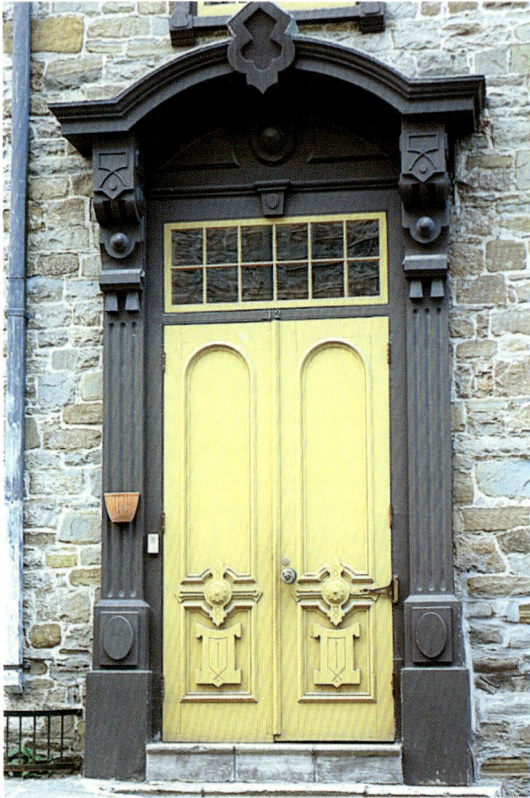

DW-80 门斗：带装饰壁柱的拱，雕刻镶板门，十二块玻璃板
门顶窗，19世纪建筑

Door hood: arch with ornate pilasters, carved paneled doors,
twelve-light transom panel; 19th-c. building

摄影：Austin, Patterson, Disston

DW-81 托斯卡纳式风格喷漆木柱和带三角檐饰门廊的古典柱
头，镶板门，菱形边窗，传统房屋

Tuscan-style painted wood columns and entablature with
pediment porch, paneled door, diamond side lights, traditional
house

摄影：Austin, Patterson, Disston

DW-82 窗上带有窗楣的半木结构的喷漆拉毛水泥木质车辆门
道，杂色的佛蒙特石板瓦屋顶

Half-timber-style stucco and painted wood porte-cochère with
eyebrow lintel over window, variegated Vermont slate roof

DW-83 新古典式山墙形门斗在壁柱和拱形门框上方；有扇形
窗的木板门

Neoclassical pedimented door hood over pilasters and arched
doorframe; wood panel door with fanlight

入口

门廊

摄影：Austin, Patterson, Disston

DW-84 门斗安在附墙支柱和带有菱形窗格条的双扇门上方，传统房屋
Door hood over engaged columns and double doors with diamond muntins, traditional house

DW-85 边缘有装饰托架的冠顶饰门斗，托架位于带有边窗的镶板门上方，19世纪美国房屋
Crown molding door hood on cut brackets over paneled door with side lights, 19th-c. American house

DW-86 有冠顶饰的拱形入口，边窗和扇形窗四周有附壁柱镶板门，18世纪美国房屋
Arched entry with crown molding and pilasters around paneled door, side lights, and fanlight, 18th-c. American house

入口

DW-87 有装饰性钟楼和门外围装饰的西南地区风格的入口
Southwestern-mission-style entry with decorative belfry and profiled door surround

DW-88 山形墙下有格子百叶的传统日式入口
Traditional Japanese entry with lattice screen under gable

DW-89 带有延伸的屋檐的传统日式柱梁式入口，双扇全板门，全玻璃边窗和格栅门，现代房屋
Traditional Japanese post-and-beam-style entry with extended eaves, double flush doors, full glass side lights and lattice gates, modern house

入口

DW-90 西班牙殖民风格凹式门在拉毛水泥墙面上的以装饰性
釉面砖为边框的弧形拱中，墨西哥瓷砖台阶
*Spanish-colonial–style recessed door in segmental arch framed
with decorated glazed tiles on stucco wall; Mexican tile steps*

DW-91 门斗在以锻铁花边带为边框的拱形入口上方；全玻璃
外有带装饰格栅的金属门，现代房屋
*Door hood over arched doorway framed with wrought-iron
filigree band; metal door with grille over full glass, modern house*

DW-92 西班牙殖民风格酒店入口弯拱下的凹式木门
*Spanish-colonial–style hotel entry with recessed wood gate under
cambered arch*

DW-93 有马蹄形拱和台阶的凹式入口
Recessed entry with ornate horseshoe arch and steps

入口

DW-94 拉毛粉刷的现代海滨别墅的露台入口：全玻璃式天然木门
Patio entry to stucco modern beach house: natural wood door with full glass

DW-95 现代房屋的庭院入口：有木栏的天然木板门上的门斗
Courtyard entry to modern house: door hood over natural wood paneled door with wood grille

DW-96 传统日式入口：双扇木门和石头台阶
Traditional Japanese entry: double wood doors and stone steps

入口

DW-97 金属框玻璃嵌板的门廊，现代商业大厦
Portico of metal-framed glass panels, modern commercial building

DW-98 装饰艺术风格酒店入口：被装饰陶瓷柱框住的台阶
Art Deco–style hotel entrance: steps flanked by decorated ceramic piers

DW-99 格栅包门和带壁柱的边窗，现代商业建筑
Grille-covered door and side lights with pilasters, modern commercial building

DW-100 侧门入口包着不锈钢嵌板，现代音乐大厅
Side entrance clad with stainless steel panels, modern concert hall

入口

DW-101 有彩色大玻璃窗和顶窗的入口，现代商业建筑
Entry with large colored glass window and overhead lighted glass panels, modern commercial building

DW-102 盖板门廊：嵌在玻璃嵌板墙里的全玻璃的不锈钢双扇门；砖石建筑环绕，现代商业建筑
Covered entryway: stainless steel double doors with full glass set in glass-paneled wall; masonry surround, modern commercial building

DW-103 喷漆金属双扇门，两侧为边窗，不锈钢嵌板和玻璃砖墙
Painted metal double doors flanked by side lights, stainless steel panels, and window wall

DW-104 有木质板条门的粗石弯拱入口
Rustic stone cambered arch entry with wood battened door

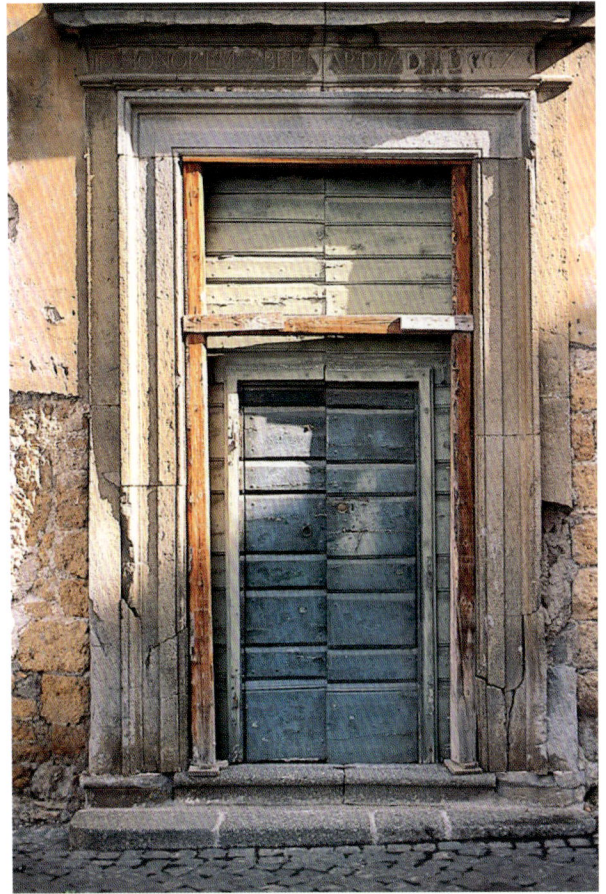

DW-105 带木质线脚、门楣镶板和喷漆镶板门的风化入口
Weathered entry with wood molding, transom panel, and painted paneled door

DW-106 成为废墟的咖啡馆入口，由带木框的窗户围绕着边窗的双扇木镶板门
Café entry in ruins: wood-paneled double doors with lights flanked by wood-framed windows

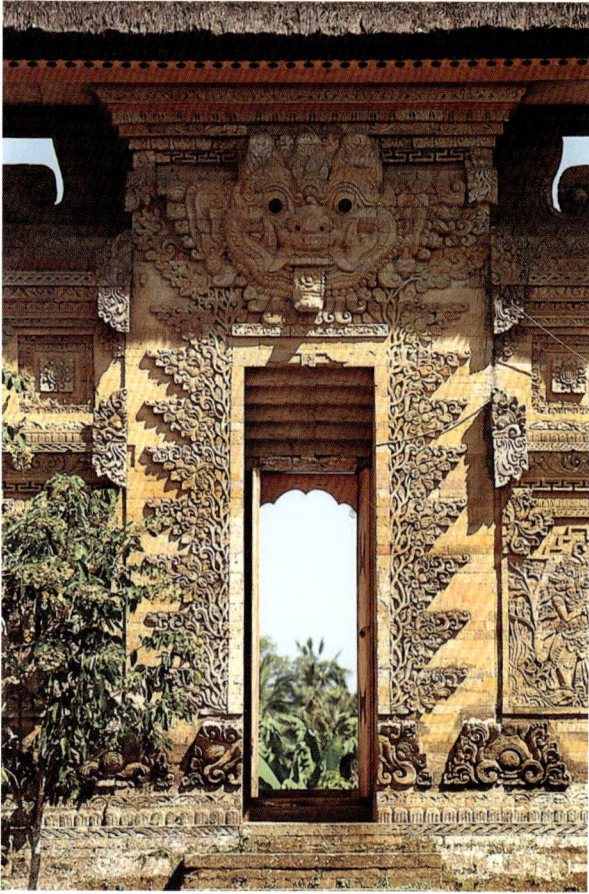

DW-107 寺庙庭院入口：石质台阶和门周围奇形怪状的雕刻，巴厘岛
Entrance to temple courtyard: stone steps and carved surround with grotesque, Bali

DW-108 中式花园的月亮门
Chinese garden moon gate

DW-109 拉毛水泥墙上带木镶板门的花园入口
Garden entrance with wood-paneled doors in a stucco wall

入口

DW-110 未抛光的木质大门，刻有凹槽的方料和粗糙打磨的原木杆交替地被钉在水平的板条上

Unfinished-wood entry gate of alternating notched square stock and roughly dressed timber posts attached to horizontal battens

DW-111 石墙面中的带有铸铜装饰的大门

Cast-bronze ornamented gate in stone wall

DW-112 由树枝垂直地固定在水平木板上组成的乡村花园大门

Rustic garden gate of vertical branches attached to horizontal battens

DW-113 粗木支柱间的木格栅滑动门，日式花园入口

Sliding wood-lattice gate between roughly dressed timber posts, entrance to Japanese garden

大门

DW-114 锻铁门嵌在墙壁内
Wrought-iron gate recessed in wall

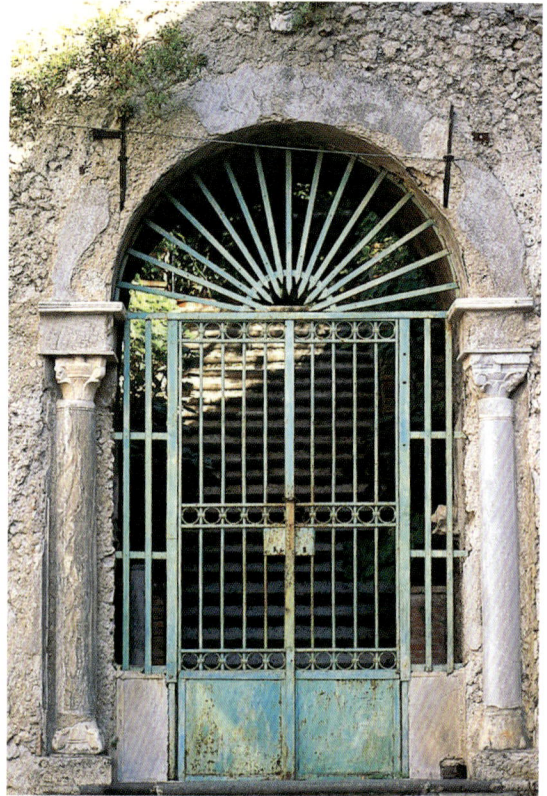

DW-115 带有开放式栅格和扇形门顶镶板的金属双扇门，门
在附墙圆柱上带有拱座的粗石拱中
Metal double gate with open bars and fan-shaped top panel in
rustic stone arch with imposts on engaged columns

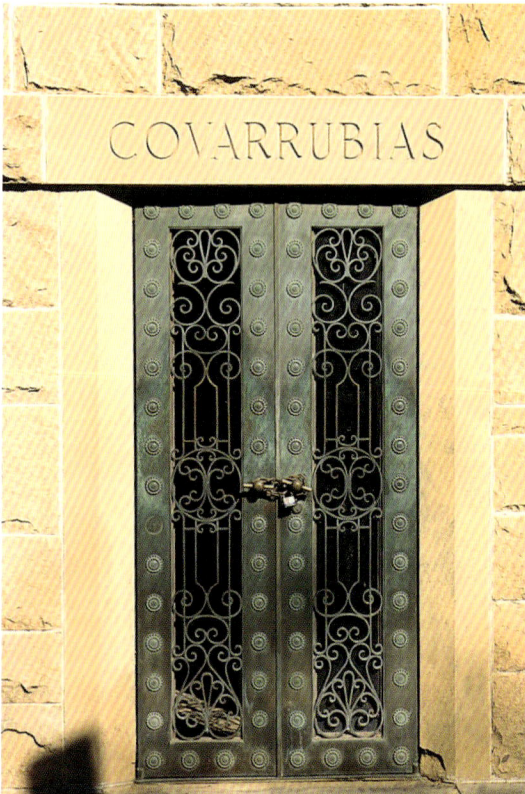

DW-116 陵墓的双扇铜门
Double bronze gates to mausoleum

DW-117 有防氧化密封护层的铸铜大门，商业建筑
Cast-bronze gate coated with sealant to prevent patination,
commercial building

大门

DW-118 上部带格栏的双扇浇铸金属门，装饰艺术风格酒店
Cast-metal double gates with upper grille, Art Deco hotel

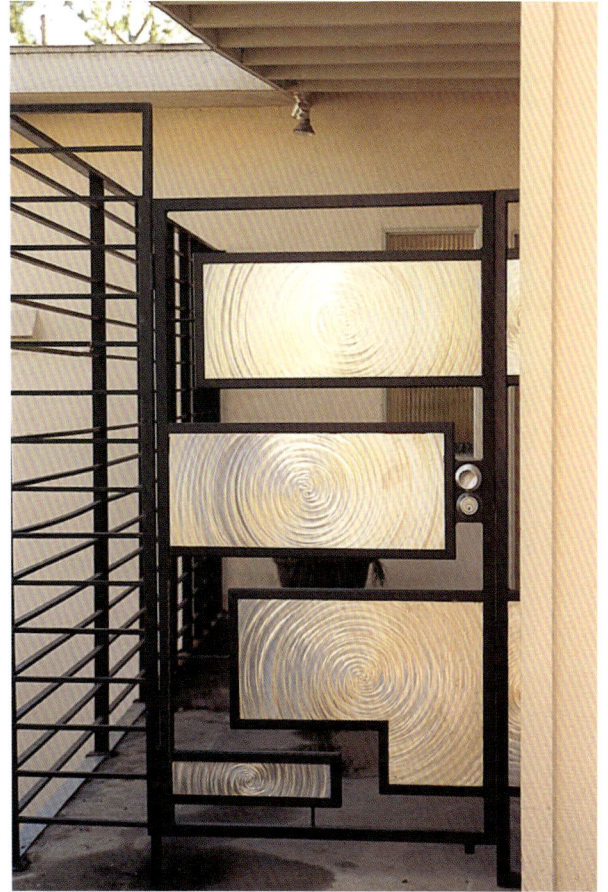

DW-119 入口处开放式矩形框中带有拉毛扩散式图案的金属镶板门，现代房屋
Entry gate of framed metal panels with brushed directional pattern in open rectangular frame, modern house

DW-120 同心环的浇涛金属入口大门，现代房屋
Entry gate of cast metal in concentric rings, modern house

大门

DW-121 有门闩和锁的浇铸的自然氧化的铜质门把手
Cast naturally patinated bronze door handle with latch and lock

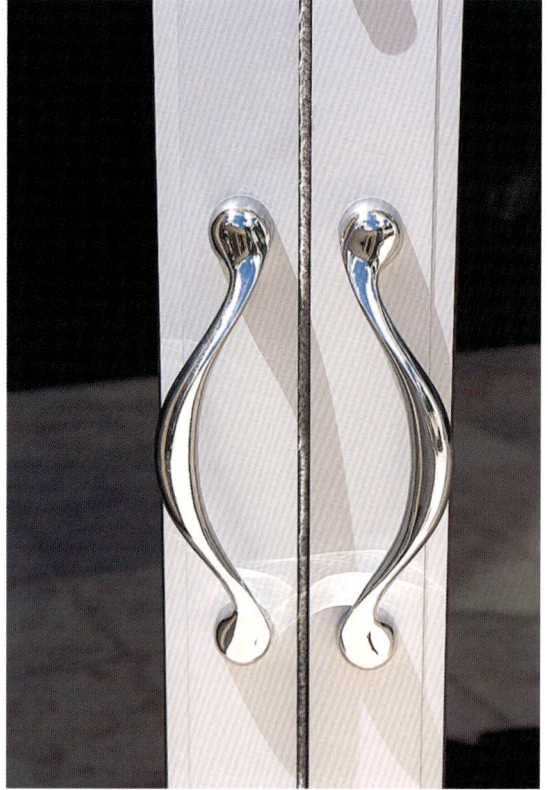

DW-122 金属双扇门中抛光的不锈钢现代风格门把手
Polished stainless steel modern door handles on metal double door

DW-123 传统日式铜质和铁质的饰件以及门拉手
Traditional Japanese bronze and iron hardware and door pull

DW-124 铸铁和黄铜的门拉手
Cast-iron and brass door-pulls

DW-125 黄铜的装饰钉、格栅和薄板
Decorative brass studs, grilles, and plates

DW-126 抛光面铜门环
Polished bronze door knocker

门饰

天花板和屋顶

摄影：Lockwood-Mathews Mansion Museum

CR-1 车道入口门道的盖有拱顶的天花板，19世纪马车入口
Vaulted ceiling of porte-cochère, 19th-c. carriage entrance

CR-2 外露的粗椽固定在土坯墙中，表面覆盖有粗锯板条；
18世纪美国西南部地区
Roughly hewn exposed rafters anchored in adobe wall and covered with rough-sawn planks; 18th-c. American Southwestern mission

CR-3 茅草屋顶表现出茅草料和框架结构之间的连接
Grass thatch roof showing connection between thatching and structure

摄影：Austin, Patterson, Disston

CR-4 裸露的天花板梁：喷漆钢框和波浪形钢板，现代房屋
Exposed ceiling beams: painted steel frame and corrugated steel decking, modern house

天花板

CR-5　粗椽后面的编织竹篾，印度尼西亚的凉亭
Woven split bamboo behind roughly hewn rafters, Indonesian pavilion

CR-6　普通住宅房屋的裸露的优质冷杉木框架
Exposed select-fir framing in residential common room

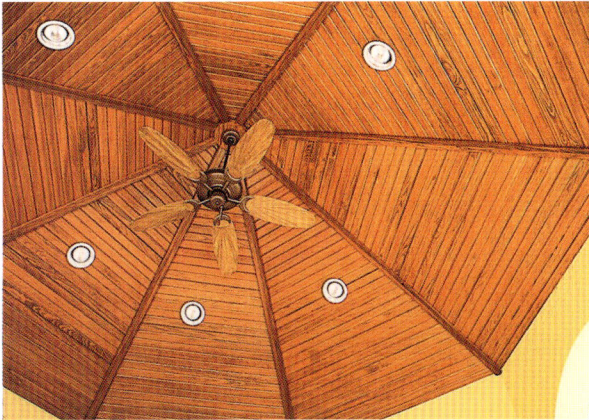

CR-7　八边形的舌榫拼接泳池天花板，现代房屋
Octagonal tongue-and-groove pool-house ceiling, modern house

CR-8　裸露的拱肋之间带有企口的喷漆半圆筒形天花板，现代房屋
Painted barrel ceiling with tongue-and-groove between exposed ribs, modern house

CR-9　拱状石膏天花板，传统房屋
Plaster vaulted ceiling, traditional house

摄影：Austin, Patterson, Disston

天花板

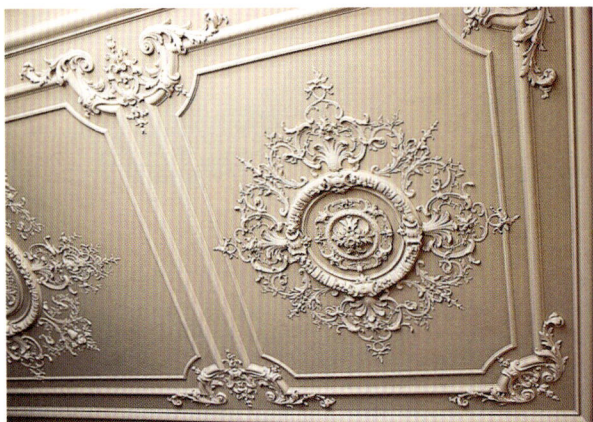

CR-10 浅浮雕装饰天花板嵌板，19世纪美国公寓
Bas-relief ornamented ceiling panel, 19th-c. American mansion

摄影：The Preservation Society of Newport County

CR-11 喷漆方格天花板，现代住宅厨房
Painted coffered ceiling, modern residential kitchen

摄影：Austin, Patterson, Disston

CR-12 相连的有装饰喷漆图案的方格天花板，19世纪美国公寓
Ceiling with decoratively painted coffers in an interlocking pattern, 19th-c. American mansion

摄影：Lockwood-Mathews Mansion Museum

CR-13 门廊上的装饰艺术风格设计中的带有对比色釉面镶板的天花板
Ceiling of contrasting colored enameled panels in Art Deco design on portico

CR-14 仿古栗木方格天花板，住宅中的图书馆
Antique-chestnut coffered ceiling, residential library

摄影：Austin, Patterson, Disston

CR-15 带有古典装饰性八边形的方格天花板
Coffered ceiling with classically ornamented octagonal coffers

天花板

CR–16 有相连的蛋形格栅的悬挂式中庭天花板，现代商业建筑
Suspended atrium ceiling with interlocking egg-crate units, modern commercial building

CR–17 带有人行天桥的金属框玻璃椭圆拱式中庭，现代商业建筑
Elliptical atrium of metal-framed glass with pedestrian bridges, modern commercial building

CR–18 带有金属框的玻璃中庭天花板，现代商业建筑
Metal-framed glass atrium ceiling, modern commercial building

CR–19 带有金属框的玻璃中庭天花板，现代商业建筑
Metal-framed glass atrium ceiling, modern commercial building

摄影：Guy Gurney

CR–20 装有玻璃的穹顶和相连的穹窿，19世纪意大利建筑
Glazed dome and adjacent vault, 19th-c. Italian building

天花板

CR-21 有板条接缝的阳极氧化铝或涂漆薄金属，现代商业建筑
Anodized aluminum or coated sheet steel with batten seams,
modern commercial building

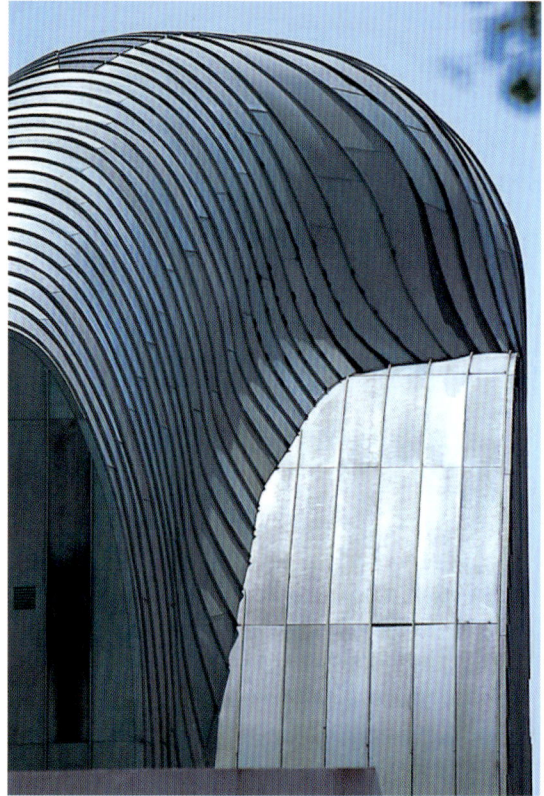

CR-22 带有接缝的锌或铜板条，现代商业建筑
Zinc or copper with flat seams, modern commercial building

CR-23 弯曲的四坡屋顶上的铜瓦
Copper shingles on curved hipped roof

CR-24 带有接缝的锌或铜板条，传统日式建筑
Zinc or copper flat seam, traditional Japanese roof

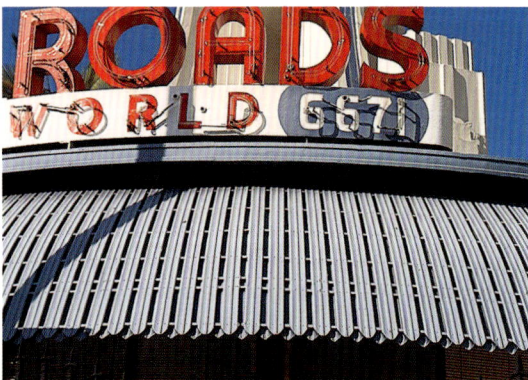

CR-25 阳极氧化铝遮檐，现代商业建筑
Anodized aluminum awning, modern commercial building

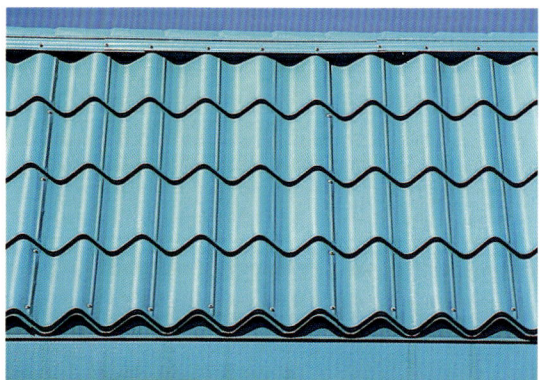

CR-26 涂有瓷釉的波浪形金属仿半圆形截面瓦
Enameled corrugated metal imitating mission tile

金属屋顶

CR-27 阳极氧化铝或立接缝的涂漆薄金属，弯曲山墙顶，现代商业建筑
Anodized aluminum or coated sheet steel with standing seams, curved gable roof, modern commercial building

CR-28 屋顶和老虎窗上有扁平接缝的薄钢瓦；底边上有管状防雪栏
Sheet-metal shingles with flat seams on roof and dormers; pipe-style snow guard on lower edge

CR-29 复折屋顶和老虎窗上有板条接缝的薄金属板；19世纪法国公寓
Sheet metal with batten seams on gambrel roof and dormers, 19th-c. French apartment house

金属屋顶

CR-30 有立接缝的阳极氧化铝或涂漆薄金属,是四坡屋顶的
一种变化,现代房屋
Anodized aluminum or coated sheet steel with standing seams,
variation of hipped roof, modern house

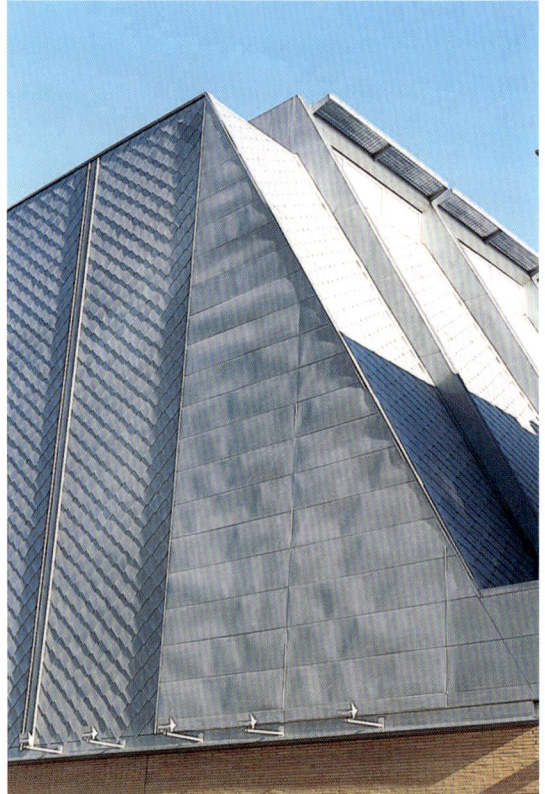

CR-31 屋顶和山墙面上有扁平或板条形接缝的阳极氧化铝或
或涂漆薄金属,现代商业建筑
Anodized aluminum or coated sheet steel with a variety of flat and
batten seams on roof and gable wall, modern commercial building

CR-32 角炮塔拱顶上的波浪状金属板,在拱顶和墙面的交界
处还有防雨板
Quilted sheet metal on dome roof of corner turret with flashing at
juncture of dome and wall

CR-33 圆锥形角楼顶带有立接缝的薄金属板;四坡屋顶和老
虎窗,19世纪法国城堡式酒店
Sheet metal with standing seams on conical turret roof; hipped
roof and hipped roof dormer, 19th-c. chateâu-style hotel

金属屋顶

CR-34 在四坡圆锥屋顶上带有板条接缝和立接缝的铜（位于顶部）及其他金属薄板，19世纪法国城堡式多功能建筑
Copper (top) and other varieties of sheet metal with batten and standing seams on hipped and conical roofs, 19th-c. chateâu-style mixed-use buildings

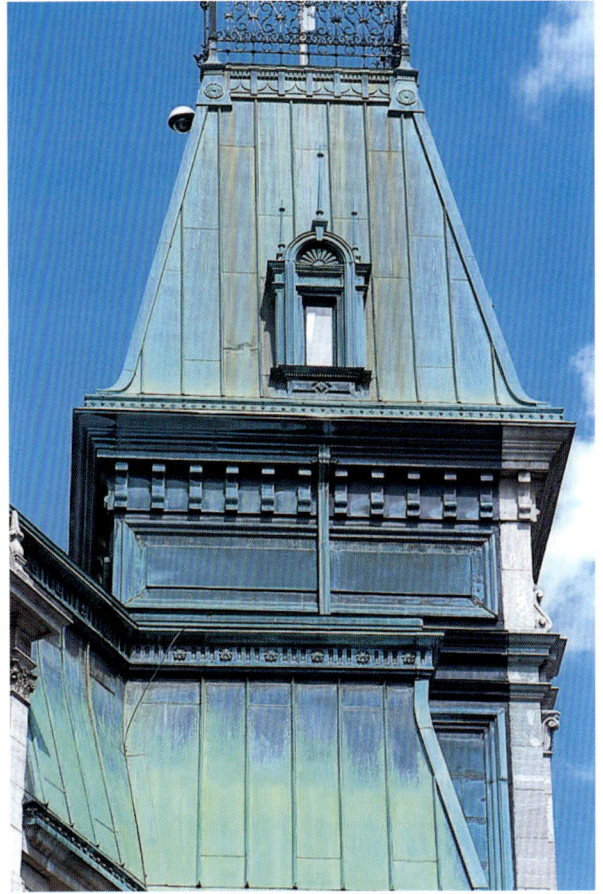

CR-35 带有立接缝的铜板，有顶饰和拱顶老虎窗的四坡屋顶；墙上有铜覆层和线脚
Copper with standing seams, hipped roof with cresting and arched roof dormer; copper cladding and molding on walls

CR-36 山墙顶和玻璃天窗有立接缝的薄金属板，现代商业建筑
Sheet metal with standing seams on gable roof and glazed monitors, modern commercial building

金属屋顶

CR-37 谷仓的喇叭形山墙上的风化波纹金属屋顶和粮仓上的圆屋顶
Weathered corrugated metal roofs with flared gable on barn and domed roof on silo

CR-38 风化波纹金属谷仓顶，天窗上有攒尖顶
Weathered corrugated metal barn roof with pyramidal roof on monitor

CR-39 有四坡顶老虎窗的立接缝板条
Standing seams with hipped roof dormer

CR-40 斜脊梁瓦端有瓦檐饰的风化陶土波形瓦，巴厘岛房屋
Weathered terracotta pantile with antefix at end of ridge tiles on hip, Balinese house

CR-41 新的和已风化的陶土筒瓦，从顶部到底部：单坡顶、山墙顶、半圆锥形顶
New and weathered terracotta mission tiles. Top to bottom: shed roof; gable roof, half-cone roof

CR-42 传统日式瓦；左侧：hogawara,右：kawara
Traditional Japanese tiles. Left: hogawara; right: kawara

CR-43 单坡顶上的风化陶土筒瓦和扁形瓦
Weathered terracotta mission and flat tiles on shed roof

瓦顶

CR-44 彩釉搪瓷尾砖，中式房顶
Polychrome glazed ceramic end-tiles, Chinese roof

摄影：Duane Langenwalter

CR-45 斜脊带有梁瓦的四坡陶土波形瓦
Terracotta pantiles on hipped roof with ridge tiles on hip

CR-46 攒尖屋顶上的风化陶土瓦
Weathered terracotta tiles on pyramidal roof

CR-47 四坡顶上的模铸混凝土波形瓦，斜脊上带有混凝土脊瓦
Cast concrete pantiles on hipped roof with concrete ridge tiles on hip

CR-48 带有屋脊饰的日式kawara瓦
Japanese kawara tiles with ornamental tile antefixes

瓦顶

摄影：Guy Gurney

CR-49 有四坡顶老虎窗的彩色扁墙面板，20世纪美国房屋
Flat multicolored shingle tiles with hipped roof dormers, 20th-c. American house

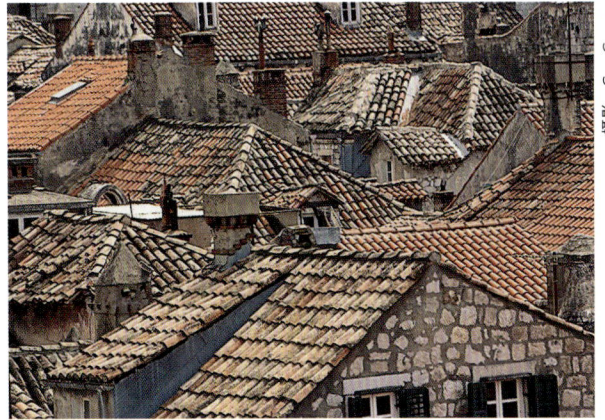

CR-50 瓦顶，亚德里海的海边村落
Tile rooftops, Adriatic coastal village

CR-51 山墙和主屋顶交叉处的筒瓦的细部
Detail of mission tiles at junction of gable and main roof

CR-52 有山墙顶老虎窗的风化扁墙面瓦
Weathered flat shingle tiles with gabled roof dormer

CR-53 有铜线脚的凸曲面弯顶上的扁墙面瓦，19世纪多功能建筑
Flat shingle tiles on convex curved roof with copper molding, 19th-c. mixed-use building

瓦顶

CR-54 带有装饰脊瓦的山墙顶，扁平陶土瓦，19世纪英国房屋
Flat shingle terracotta tiles on gable roof with decorative ridge tiles, 19th-c. English house

CR-55 四坡屋顶和窗上方单坡顶上的彩色半圆瓦，20世纪美国房屋
Multicolored mission tiles on hipped roof with shed roof over window, 20th-c. American house

CR-56 变化四坡上有边饰的陶土波形瓦，巴厘岛房屋
Terracotta pantiles with ridge ornament on variation of hipped roof, Balinese house

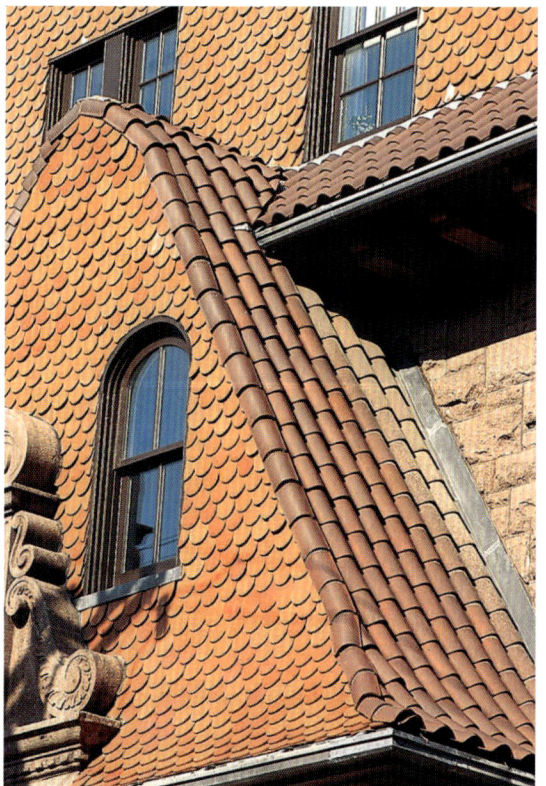

CR-57 筒瓦顶，在扇形瓦贴面板中，带有特别的山墙和外立面，20世纪美国房屋
Mission-tile roof with idiosyncratic gable and facade clad in scalloped tile shingles, 20th-c. American house

瓦顶

CR-58 有攒尖顶老虎窗的四坡顶，屋顶为杂色石板瓦，美国房屋
Variegated slate roofing on hipped roof with pyramidal wall dormer, American house

CR-59 带有三角檐饰的老虎窗四周的带状彩色图案扇形板瓦
Scalloped slate roofing in banded polychrome patterns around pedimented dormer

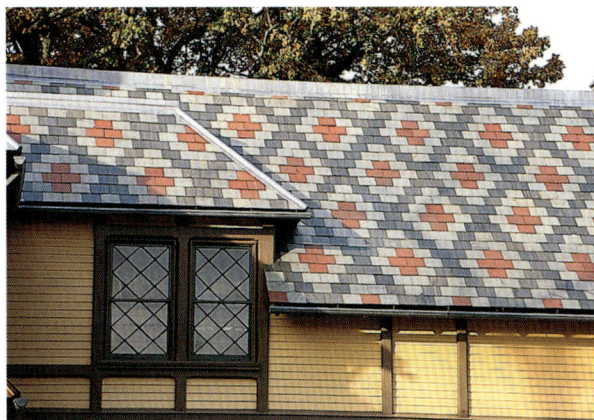

CR-60 彩色菱形图案中的长方形石板瓦
Rectangular slate roofing in polychrome diaper pattern

CR-61 带有防雪栏的杂色石板瓦
Variegated Vermont slate roofing with snow guards

摄影：Austin, Patterson, Disston

CR-62 有星星装饰的带状图案扇形石板瓦
Scalloped slate roofing in banded pattern with inset stars

CR-63 带有眉形老虎窗的杂色石板瓦
Variegated slate with eyebrow dormer

石板屋顶

CR-64 扇形和菱形喷漆木瓦
Scalloped and diamond-shaped painted wood shingles

摄影：Austin, Patterson, Disston

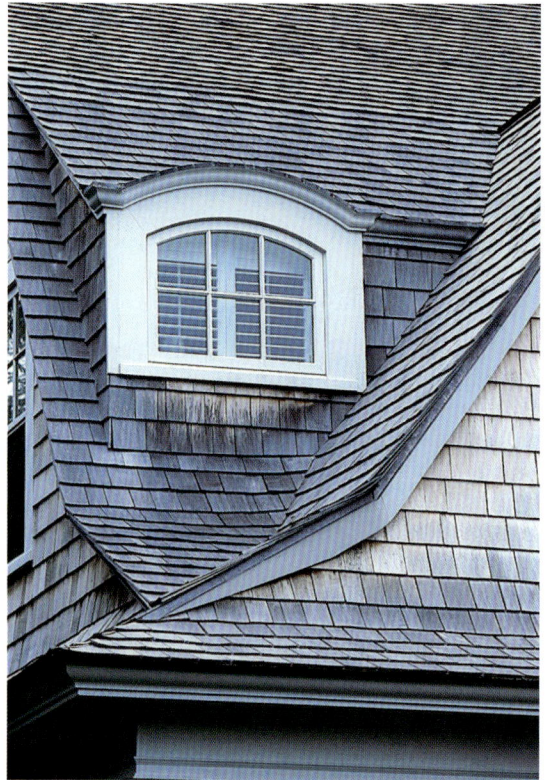

CR-65 阿拉斯加黄杉木瓦；山墙相交处为屋顶排水沟；传统美国房屋
Alaskan yellow cedar shingles; roof valley at intersection of gables; traditional American house.

CR-66 山墙顶的竹顶，印度尼西亚
Split bamboo shingles on gabled roof, Indonesian house

CR-67 四坡屋顶上带有瓦檐饰的木瓦，巴厘岛神龛
Wood shingles on hipped roof with antefix, Balinese shrine

CR-68 带有拱顶老虎窗的屋顶上的石油沥青瓦
Asphalt shingles on roof with arched wall dormers

CR-69 四坡屋顶上的木瓦，沿着屋脊还有为对比色的木瓦
Wood shingles on hipped roof with contrasting shingles along hip

木瓦屋顶

CR-70 砖混建筑中的装饰茅草屋顶，巴厘岛
Trimmed thatch on concrete-block building, Bali

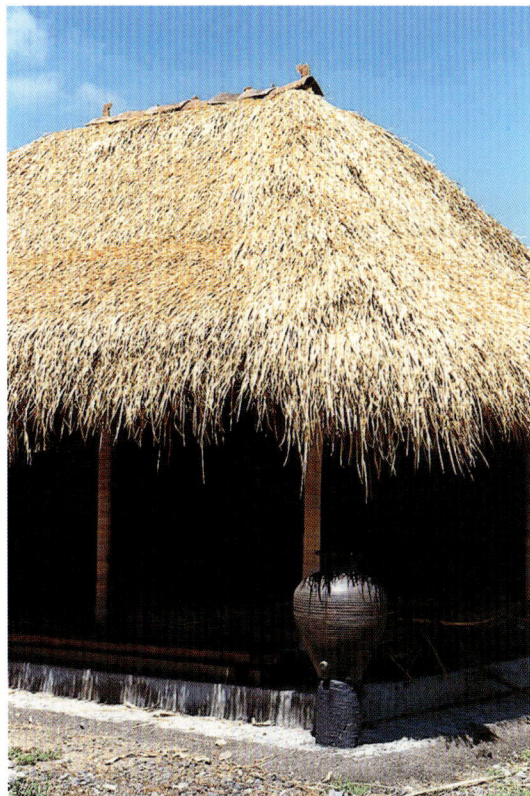

CR-71 棕榈叶屋顶，巴厘岛的亭子
Palm-frond thatch, Balinese pavilion

CR-72 传统棕榈叶屋顶亭子的下表面
Underside of traditional palm thatched pavilion

CR-73 传统英国茅草顶房屋上的扇形顶边
Scalloped roof ridge on traditional English thatched house

CR-74 山墙四周的茅草顶细部，英国谷仓
Detail of thatching around gable, English barn

CR-75 房屋上的茅草屋顶，拉丁美洲
Thatched roof on house, Latin America

摄影：Meredith Barchat

草屋顶

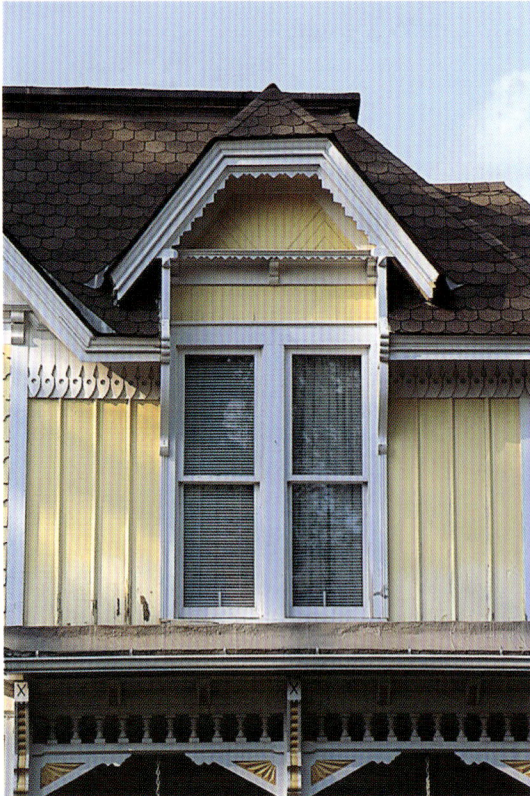

CR-76 有两坡式屋顶的墙面屋顶窗
Wall dormer with jerkinhead roof

CR-77 有瓦顶的半木风格山墙，20世纪美国房屋
Half-timber–style gable wall with tile roof, 20th-c. American house

摄影：Guy Gurney

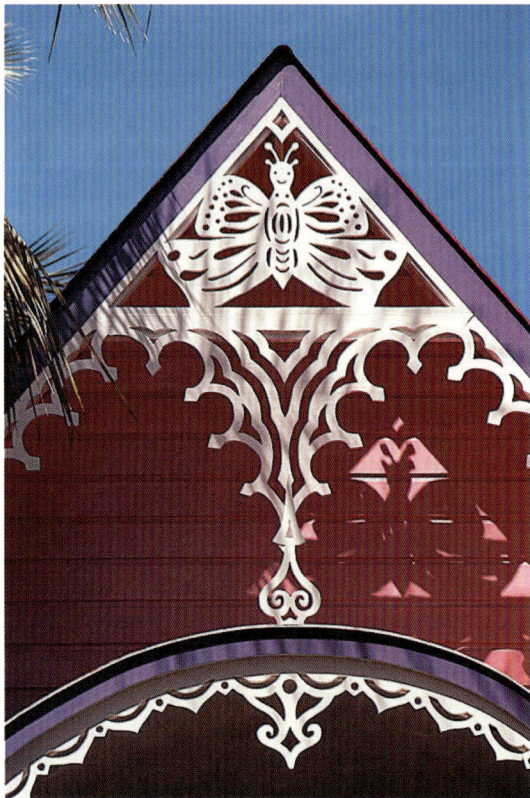

CR-78 镂空的鲜艳华丽风格的山墙和山墙封檐板装饰，加勒比地区房屋
Pierced gingerbread-style gable and bargeboard ornament, Caribbean house

CR-79 有雕刻的半木风格封檐板的山墙面，19世纪美国房屋
Gable wall with carved half-timber–style bargeboard, 19th-c. American house

山墙与屋顶窗

CR-80 有装饰山墙封檐板的墙面屋顶，美国墙面板和木瓦房屋
Wall dormer with decorated bargeboard, American clapboard and shingle house

CR-81 有巴拉迪欧式窗扇的屋顶窗
Roof dormer with Palladian window

摄影：Austin, Patterson, Disston

CR-82 覆盖有石板瓦的双屋顶窗
Double roof dormers clad with slate shingles

摄影：Austin, Patterson, Disston

CR-83 墙面屋顶窗，左侧：木瓦贴面，右侧：木雕饰，美国房屋
Wall dormers. Left: shingle-clad; right: wood with carved ornament, American house

CR-84 带有格栅的木质山墙；山墙顶和毗邻的单坡顶（底部）为陶瓦，日本寺庙
Wood gable with lattice; ceramic tiles on gable roof and adjoining shed roof (bottom), Japanese temple

CR-85 有鲜艳华丽风格装饰的山墙封檐板的双墙面屋顶窗，19世纪美国石砌房屋
Double wall dormers with gingerbread-style decorated bargeboard, 19th-c. American stone house

CR-86 喇叭形屋檐的门廊上的木瓦屋顶窗；石头山墙端墙面
Shingled roof dormer above flared eave covering porch; stone gable-end wall

CR-87 带有筒瓦的两坡式日屋顶，四坡顶屋顶窗，山墙端墙面
Dormer with hipped roof on jerkinhead tile roof with scalloped terracotta tiles, gable-end wall

CR-88 风化线脚上有新平线脚的山墙端墙面；带攒角顶的玻璃天窗
Gable-end wall with new flush board siding above weathered siding; glazed monitor with pyramidal roof

山墙与屋顶窗

CR-89 喷漆木顶上的墙面天窗
Wall dormer on painted wood roof

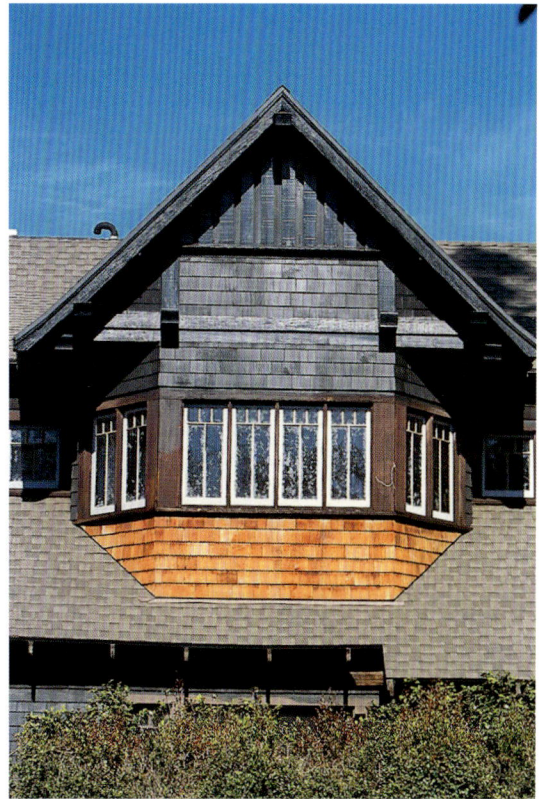

CR-90 有窗户和木梁的跨越式山墙顶的三边顶屋顶窗，艺术
手工杉木板房
Three-sided roof dormer with windows and wood beam spanning
gable roof, Arts and Crafts cedar-shingle house

CR-91 四坡山墙顶屋顶窗；屋顶窗和屋顶有波形瓦覆盖
Hipped gable roof dormer; dormer and roof clad with pantiles

CR-92 有巴拉迪欧式窗扇的山墙面屋顶窗
Gabled wall dormer with Palladian window

山墙与屋顶窗

CR-93 带有三角檐饰的屋顶和单坡顶的屋顶窗
Pedimented gable roof dormer and shed-roof dormer

CR-94 个性化的半木结构山墙，16世纪英国房屋
Idiosyncratic half-timber gable, 16th-c. English house

CR-95 半木结构山墙，带有扇形山墙封檐板的单坡顶，屋顶排水沟
有防雨板瓦
Half-timber wall gables with scalloped bargeboard flanking shed roof and
flashing in roof valleys

CR-96 包有杉木瓦的渐宽式复折屋顶的山墙端墙面；枕梁瓦跨越式
山墙
Gable-end wall of flared gambrel roof with white cedar shingles; corbelled
tie spanning gable

CR-97 三扇四坡顶屋顶窗和一个带有山墙形屋顶窗；屋顶窗和屋顶
上为石板瓦
Three hipped-roof dormers and a gabled wall dormer; slate cladding on
roof dormers and roofs

CR-98 带有金属屋顶板的复折式屋顶，带有半圆拱顶的屋顶窗
Segmental-arch roof dormers on mansard roof with metal shingles

山墙与屋顶窗

CR-99 金属屋顶上的平顶墙面屋顶窗
Flat-roofed wall dormers on metal roof

CR-100 喇叭形攒尖顶上的四扇喇叭形山墙顶屋顶窗
Four flared-gable roof dormers above flared pyramidal roof dormers on metal hipped roof

CR-101 有立接缝的金属四坡顶上的单坡顶屋顶窗，现代房屋
Shed-roof dormer on metal hipped roof with standing seams, modern house

CR-102 带有防雪栏的金属四坡屋顶上的四坡顶屋顶窗，现代公寓
Hipped-roof dormers on metal hipped roof with pipe snow guards, modern apartment house

CR-103 有深屋檐的山墙形屋顶窗
Gable-roof dormers with deep eaves

CR-104 复折式屋顶上的单坡顶屋顶窗
Shed-roof dormers on gambrel roof

山墙与屋顶窗

CR-105 木质结构中的悬臂式屋檐，传统日式宝塔
Cantilevered eave over wood structure, traditional Japanese pagoda

CR-106 屋檐侧托架
Profiled eave brackets

CR-107 木屋檐托架支撑着筒瓦四坡屋顶，美国西南部地区风格的房屋
Wood eave brackets supporting mission-tile hipped roof, American Southwestern mission–style house

CR-108 现代地中海式风格房屋中的屋檐侧托架
Profiled eave brackets on modern Mediterranean-style house

CR-109 多层平顶的突悬体，现代房屋
Overhangs of multileveled flat roofs, modern house

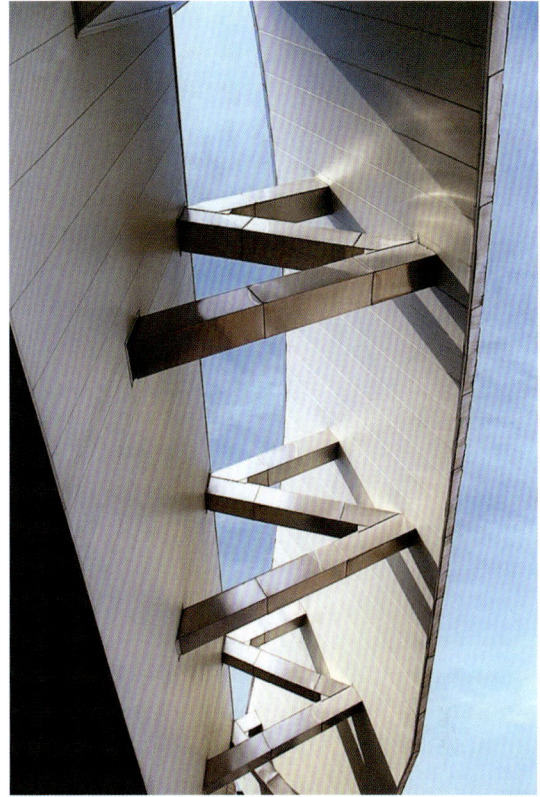

CR-110 用Z字形托架连在外立面上的金属屋顶的半独立式延伸部分，现代商业建筑
Semidetached extension of metal roof connected to facade by Z-shaped brackets, modern commercial building

CR-111 V形切口叉梁上的四坡屋顶的屋檐，艺术手工房屋
Eaves of hipped roof on notched cross-beams, Arts and Crafts house

CR-112 以金属檐作为门斗的金属和木头山墙，现代海滨别墅
Metal and wood gable with eave acting as door hood, modern beach house

CR-113 四坡顶屋顶窗和山墙形屋顶窗上的多层屋檐
Multiple eaves on hipped roof dormers and gable-wall dormers

CR-114 单坡顶屋檐连接着阳台和有遮檐的庭院门廊，传统日式房屋
Eave of shed roof connected to balcony to shelter courtyard walkway, traditional Japanese house

屋檐

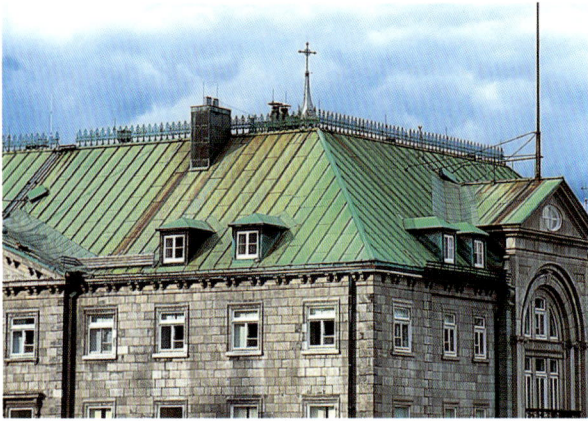

CR-115 带有立接缝的四坡铜屋顶；四坡顶屋顶窗（左）和山墙形屋顶窗（右）

Hipped, copper with standing seams; hipped-roof dormers (left) and pedimented wall dormer (right)

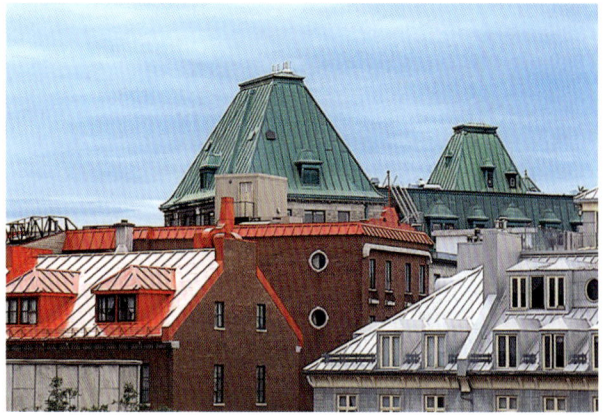

CR-116 回坡顶屋顶窗的如墙（左），有屋脊天窗的四坡屋顶（右），有屋脊天窗的喇叭形复斜屋顶（顶部），全部为薄金属板

Parapeted gable with hipped-roof dormers (left), hipped roof with hip dormers (right), and flared mansard with hip dormers (top), all sheet metal

摄影：Roger Bartels Architects, Dobyan & Dobyan Builders

CR-117 有单坡顶屋顶窗和单坡顶凸窗的喇叭形山墙板屋顶

Flared-gable shingle roof with shed dormer and shed bay window

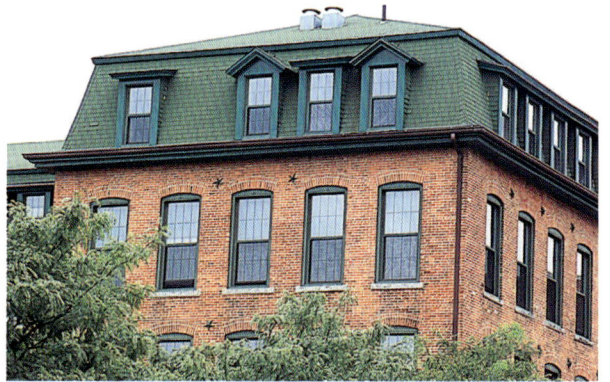

CR-118 有单坡顶屋顶窗的复折式屋顶（右），交替排列的山墙形和单坡屋顶窗

Mansard with row of shed dormers (right) and alternating gable and shed dormers (left)

CR-119 左：有山墙天窗和塔楼的山墙；右：装有风化风向标的攒尖屋顶

Left: gable with gable monitor and cupola; right: pyramid roof with weathervane

屋顶种类

CR-120 左：有屋顶窗的山墙和平接缝的金属板；右：有四坡顶屋顶窗的山墙，木瓦；屋顶排水沟有防雨板
Left: gable with roof dormers, metal with flat seam; right: gable with hipped-roof dormers, wood shingle; flashing in roof valley

CR-121 带有木瓦墙面板的斜盖盐箱形房屋
Shingled saltbox

CR-122 圆形屋顶，有檐口托饰的窗过梁饰带
Compass roof, fascia with modillions

CR-123 半木结构山墙，立面拱形装饰
Half-timbered gable, with surface arch

CR-124 有单坡顶屋顶窗的陡峭的半木结构山墙
Steep half-timbered gable with shed dormer

CR-125 半木结构山墙
Half-timbered gable

天花板和屋顶

CR-126 石板瓦山墙屋顶和有凸窗的山墙形屋顶窗
Slate gable roof and gabled roof dormer with bay

CR-127 带有彩色石板顶的半木风格交叉山墙
Half-timber–style cross gable with variegated slate roofing

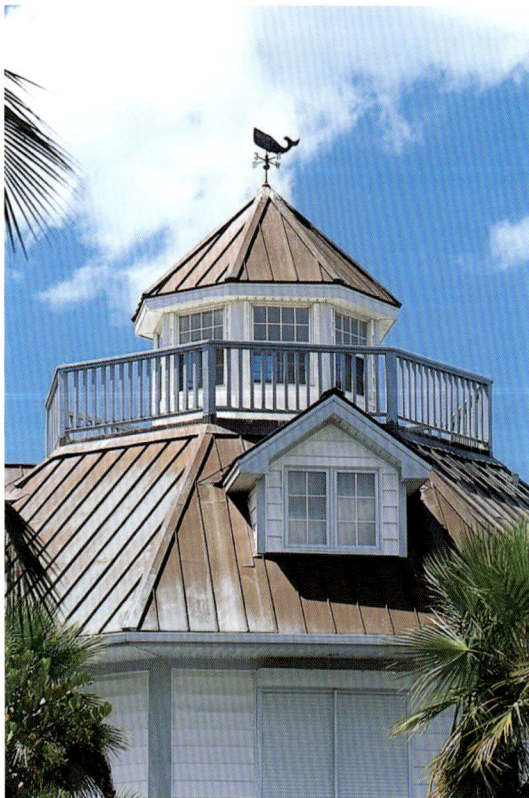

CR-128 有屋面金属板和山墙顶老虎窗的八边形喷漆立接缝金属屋顶
Octagonal painted standing-seam metal roof with widow's walk and gabled roof dormer

CR-129 渐宽的复折式屋顶，拱腹有飞檐托饰
Flared gambrel, soffit with modillions

摄影：Austin, Patterson, Disston

屋顶种类

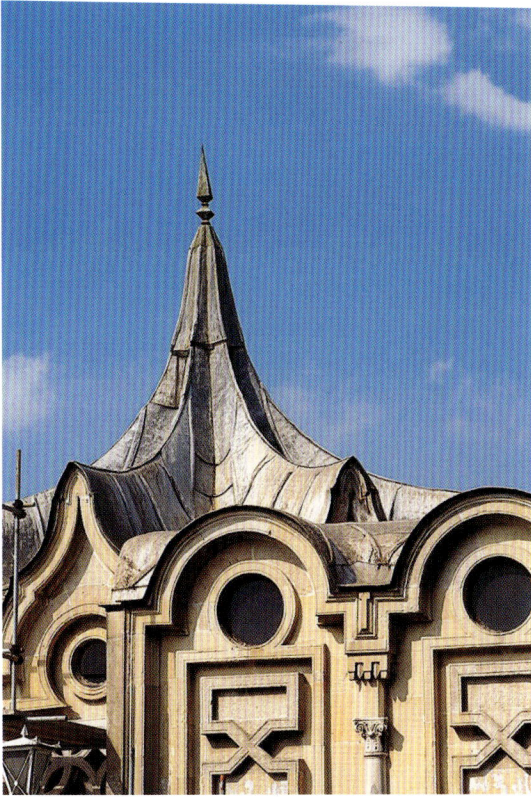

CR-130 带立接缝的个性金属屋顶
Idiosyncratic metal roof with standing seams

CR-131 复合圆屋顶，屋顶材质为有立接缝的金属
Composite dome, metal with standing-seam

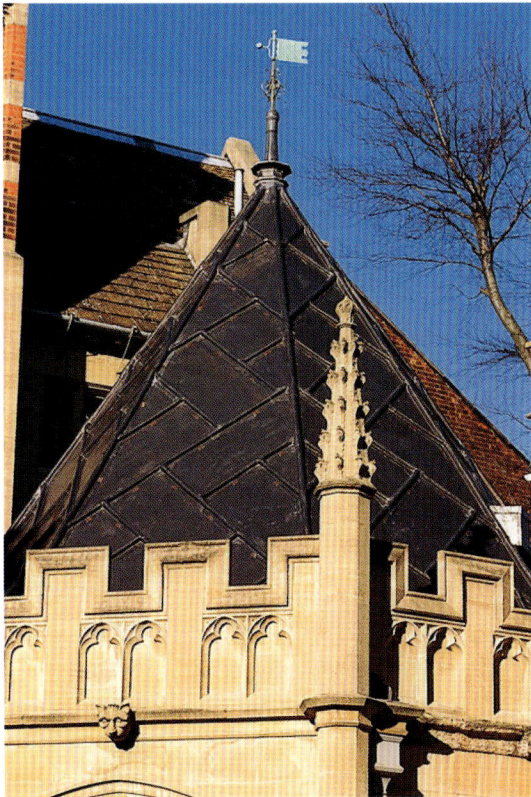

CR-132 金属攒角式屋顶
Pyramidal metal roof

CR-133 角楼上的八边形攒角屋顶，交叉山墙形屋顶窗
Pyramidal octagonal roof on turret, cross-gable dormers

屋顶种类

CR-134 S形石板瓦屋顶邻着山墙顶，都带有金属梁装饰
Ogee slate roof adjacent to gable roof, both with metal ridge ornamentation

摄影：Guy Gurney

CR-135 宝塔
Pagoda

摄影：Guy Gurney

CR-136 四坡屋顶的茅草山墙，日本寺庙
Thatched gable on hipped roof, Japanese temple

屋顶种类

CR-137 四坡屋顶：开放形屋顶板（左）;有立接缝的金属板（右）
Hipped roofs: open shingled (left); metal with standing seams (right)

CR-138 扁平屋和单坡屋顶，带有板条接缝的喷漆金属板
Shed and flat roofs, painted metal, with batten seams

CR-139 带有八边形圆顶的四坡形屋顶，带立接缝的金属板
Hipped roof with octagonal cupolas, metal with standing seams

屋顶种类

CR-140 扁平带孔屋顶，带有突出的屋檐和高过混凝土遮篷的椰子树
Flat, punctured roof with projecting eave and palm trees growing through concrete canopy

CR-141 悬臂式结构遮阳篷延伸自独立的拉毛水泥墙和天花板，户外公共空间
Cantilevered fabric awning springing from freestanding stucco wall and ceiling, outdoor public space

CR-142 有裸露的横梁、支撑和托梁的单坡顶顶入口
Shed-roofed entry with revealed rafters, braces, and joists

CR-143 砖壁柱式烟囱：螺旋形（左）；带状（右）
Brick-pilaster chimneys: spiral (left); banded (right)

CR-144 带状碎石烟囱
Coursed-rubble chimney

CR-145 有屋脊加盖的烟囱
Brick chimneys with hipped chimney hoods

CR-146 方形附墙烟囱
Engaged rectangular chimney stack

摄影：Guy Gurney

CR-147 拉毛水泥烟囱
Stucco chimney

CR-148 带有攒角顶的砖面图游拉毛粉刷的烟囱
Stucco-over-brick chimney with pyramidal hood

CR-149 砖砌烟囱
Brick chimney

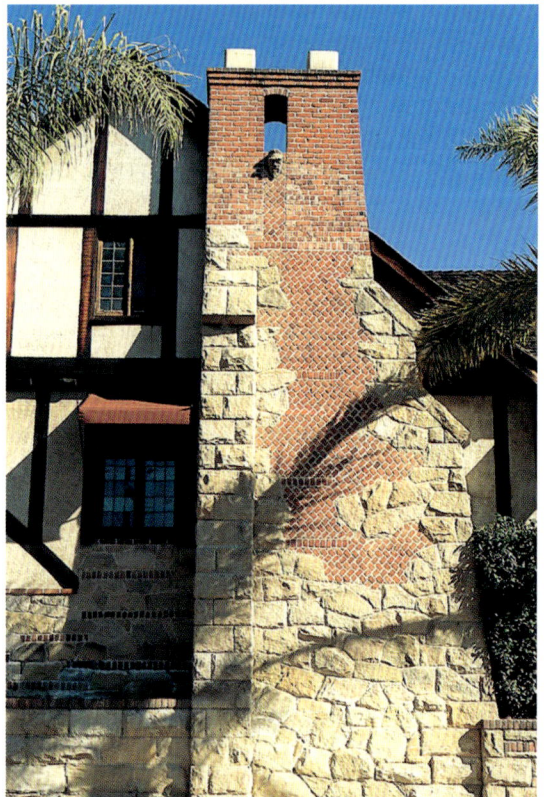

CR-150 烟囱：层砌琢石墙面扶壁和随意堆砌的琢石面，人字形图案砖墙
Chimney: coursed-ashlar buttress and random ashlar stonework, herringbone brick

烟囱

CR-151 有八边形陶制烟口的砖砌烟囱
Brick chimney with octagonal terracotta chimney pots

摄影：Austin, Patterson, Disston

CR-152 扁平接缝的金属烟囱位于对角线扁平接缝的金属屋顶上
Flat-seamed metal chimney on diagonal flat-seamed metal roofing

CR-153 层砌碎石烟囱
Coursed-rubble chimney

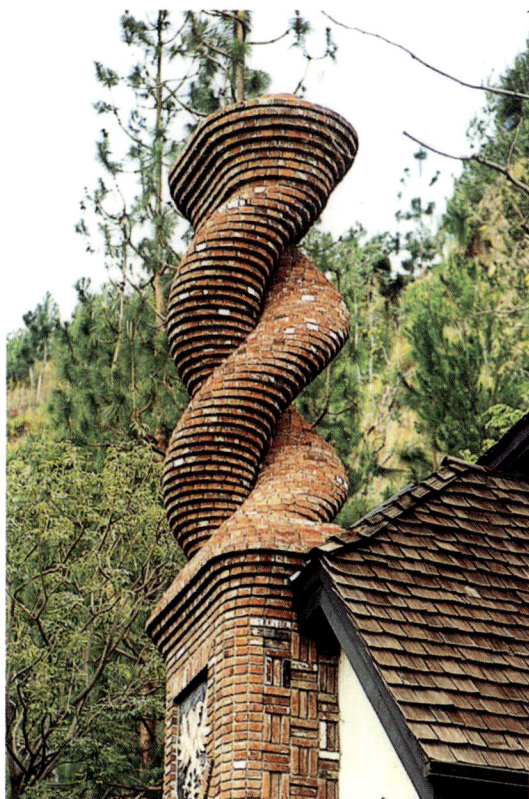

CR-154 螺旋形砖砌烟囱口
Spiral brick chimney pot

烟囱

CR-155 竹木雨水口和水落斗细部，日本
Detail of wood and bamboo gutter and rainwater head, Japan

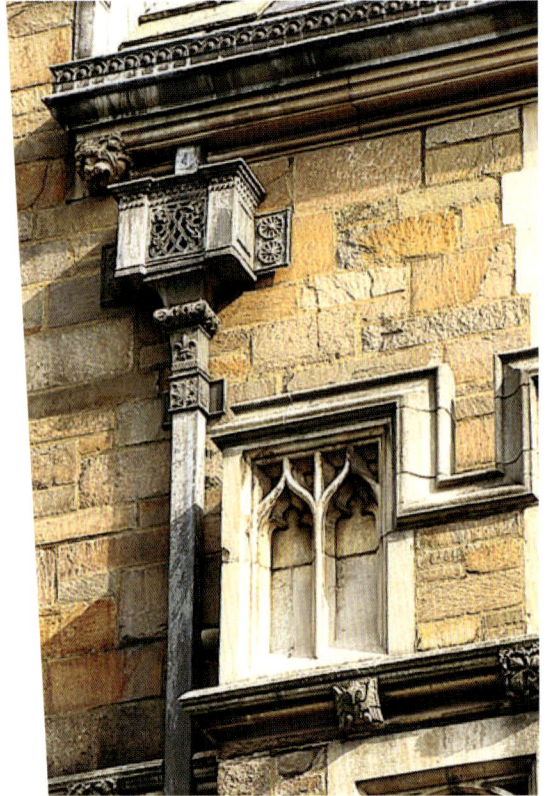

CR-156 金属装饰雨水口、水落斗和排水管
Metal decorated gutter, rainwater head, and drainpipe

CR-157 金属雨水口、排水管和水落斗
Metal gutter, drainpipe, and rainwater head

CR-158 铜排水管和水落斗
Copper drainpipe and rainwater head

雨水口与水落斗

地板和铺装

FP-1 铺面：圆形设计中的两种木料
Decking: two varieties of wood in circular design

FP-2 带有径切厚木板和木楔结构的刨光原木，乡村花园桥
Dressed logs with center-sawn planks and wood-peg construction, rustic garden bridge

FP-3 木铺面：水池环绕着内置的圆形结构
Wood decking: pool surround with inset circle

木地板与铺地

FP-4　槐木地板和中国花岗岩砖，住宅门厅
Ash flooring and Chinese granite tile, residential foyer

FP-5　从顶部到底部：木框活动地板；门槛；锯板上排列紧密的窄木片
Top to bottom: framed-wood raised floor; door sill; closely spaced narrow strips of wood over sawn planks

FP-6　带有樱桃木柜橱的加拿大榆木地板
Canadian elm flooring with cherry cabinetwork

FP-7　槐木的地板和阳台走廊的圆突沿台阶
Ash flooring and steps with rounded nosing in balcony hallway

摄影：Austin, Patterson, Disston

木地板与铺地

FP-8　有槽的风化处理木铺面和金格栅格交替排列，城市公共空间
Grooved weather-treated wood decking alternating with metal grates, urban public space

FP-9　劈开的竹铺面
Split-bamboo deck

FP-10　混凝土环绕的木板，庭院通道
Wood planks with concrete surround, courtyard walkway

FP-11　反光水池（左侧）四周的木铺面
Wood deck around reflecting pool (at left)

FP-12　石板步道上的合成木铺面，喷漆的红木长凳
Ipe-wood deck in flagstone walk; painted mahogany benches

摄影：Austin, Patterson, Disston

木地板与铺地

FP-13　门厅中有边的框六边形图案的大理石拼花地板
Marble parquet in framed hexagon pattern in foyer

FP-14　圆形设计中的大理石和化石拼花地板，休息室地板
Marble and fossil-stone parquet in circular design, pavilion floor

FP-15　门厅内的彩色几何图案大理石拼花地板
Marble parquet in polychrome geometric pattern in foyer

FP-16　门厅内的圆形放射式设计的大理石拼花地板
Marble parquet in circular radiating design in foye

FP-17　彩色几何图案中的大理石拼花地板，现代商业建筑内部
Marble parquet in polychrome geometric pattern, interior of modern commercial building

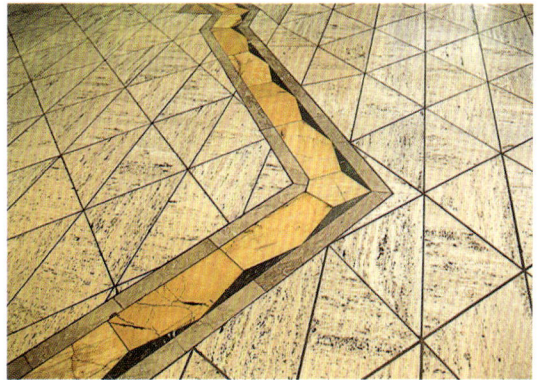

FP-18　门厅中带有彩色大理石带的几何图案洞石和大理石拼花地板
Travertine and marble parquet in geometric pattern with marble polychrome band in foyer

石头地面与铺装

FP-19 带状图案的洞石铺路石，城市公共空间
Travertine paving stone in banded pattern, urban public space

FP-20 带有立砌镶边和石水槽的顺砖砌合的规格石料，花园
通道
Dimension fieldstone in running bond with soldier-course border
and stone gutter, garden walkway

FP-21 天然石墙旁边有混凝土垫层的铺地石板，带有立砌镶
边和顺砖砌合，花园小径
Paving stones set in concrete in running bond with soldier-course
border adjacent to natural rock wall, garden path

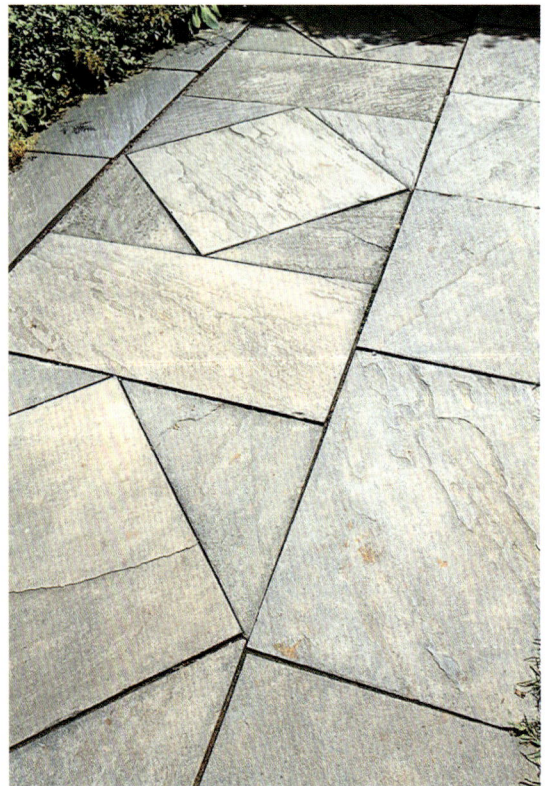

FP-22 几何图案中的规格板岩，花园小径
Dimension slate in geometric pattern, garden path

石头地面与铺装

FP-23 斜纹设计的深色和浅色的标准花岗岩铺路石，城市公共空间
Dark and light dimension-granite paving stone in diagonal design, urban public space

FP-24 多种颜色的楔方形蓝灰砂岩，人行道
Grouted squared bluestone in various colors, sidewalk

FP-25 弯曲通道上的方形面起纹铺路石，城市公共空间
Squared and roughly squared textured paving stones on curved walkway, urban public space

FP-26 被石头坡道分开的石台阶，城市公共空间
Stone stairway divided by stone ramps, urban public space

FP-27 直线接合(顶部)和顺砖砌合（底部）的起纹标准铺路石通缝铺，金属格栅和河卵石，现代商业建筑庭院
Textured dimension paving stone with straight joints (top), and running bond (bottom), metal grates, and river rocks in courtyard of modern commercial building

石头地面与铺装

FP-28　人行道上乱砌图案中的粗面方形铺路石
Roughly squared paving stone in random-ashlar pattern on sidewalk

FP-29　人行道上长苔藓的乱砌图案中的粗面方形铺路石
Roughly squared paving stone in random-ashlar pattern with moss on sidewalk

FP-30　从左到右：顺砖砌合中的铺路石，天然石头，混凝土
Left to right: paving stone in running bond; natural stones; concrete

石头地面与铺装

FP-31　街道上的铺路石
Paving stone on street

FP-32　街道上风化的铺路石
Weathered paving stone on street

FP-33　仿古面大理石铺路石，现代商业建筑的庭院
Tumbled paving stone, courtyard of modern commercial building

FP-34　斜坡状台阶中的铺路石，现代商业建筑入口
Paving stone in ramp-like steps, entry of modern commercial building

FP-35　一条古老的街道上的有趣的顺砖砌合中的大块圆石
Jumbo cobblestones in intersecting running bonds on an ancient street

FP-36　有排水口的扇形图案的铺路石
Paving stone in fan pattern with drain cover

石头地面与铺装

FP-37 顺砖砌合的铺路石（左），切开的大圆石四周带有切开的粗石；粗面方形
铺路石镶边
Paving stone in running bond (left), with split fieldstones around split bolders; border of roughly squared paving stone

FP-38 顺砖砌合的砖车道，错缝粗石被用作分隔线，艺术手工庭院
Uncoursed fieldstone used as divider in running-bond brick driveway, Arts and Crafts residence

FP-39 随意排列的圆形鹅卵石嵌在混凝土中，花园小径
Random round cobbles set in concrete, garden path

石头地面与铺装

FP-40 马赛克图案中的色彩斑驳的石板，城市公共空间
Variegated flagstone in mosaic pattern, urban public space

FP-41 顺砖砌合的方形起纹铺路石之间的切开的光面大圆石，城市公共空间
Split, polished boulders between squared and textured paving stones in running bond, urban public space

FP-42 镶在混凝土中的切开的光面小粗石和浅色石板镶边，形成了台阶
Small, split, polished fieldstones set in concrete and bordering light-colored flagstone, forming steps

石头地面与铺装

FP-43 被切开的天然石头包围着的圆形大圆石嵌在混凝土中，花园小径
Large round boulders surrounded by split natural stones, both set in concrete, garden path

FP-44 带状图案中的粗面带纹理方石，花园人行道
Roughly squared, textured dimension stone in banded pattern, garden walk

FP-45 斑驳的石板，灵活的铺设，露台地面
Variegated flagstone; flexible paving, patio floor

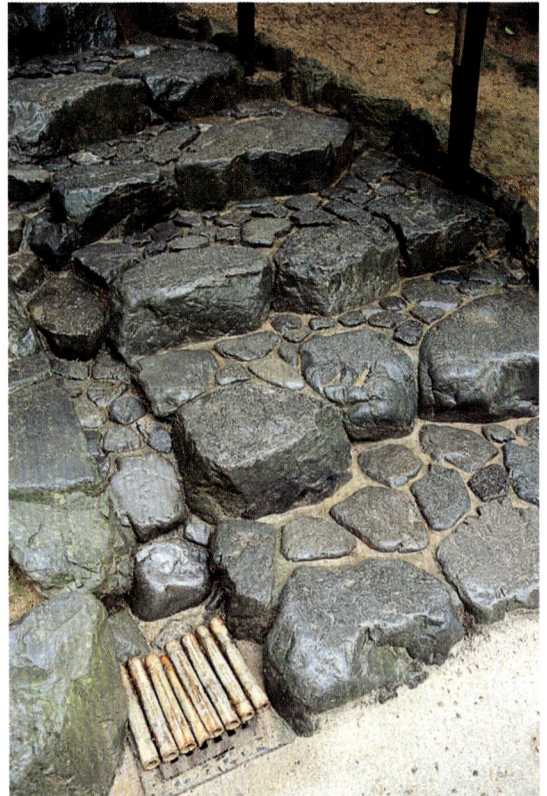

FP-46 粗面方形切割的嵌在混凝土中的光面粗石构成的台阶，左下：竹质排水盖
Steps made from roughly squared split, polished fieldstone set in concrete. Bottom left: bamboo drain cover

石头地面与铺装

FP-47　岩石花园中的石板路
Flagstone path in rock garden

FP-48　花园小径上的带有分割石嵌饰的方形花岗岩
Squared granite blocks with split-stone inset on garden path

FP-49　池塘的圆柱形踏脚石，日本公园
Cylindrical steppingstones in pond, Japanesegarden

FP-50　带有边饰的菱形图案的粗面方形铺装，花园小径
Roughly squared paving stone in diamond pattern with border, garden path

石头地面与铺装

摄影：Austin, Patterson, Disston

FP-51 混凝土浴缸四周带有排水孔的锤琢石灰岩和卵石，浴室
Hammered limestone and beach pebbles with drain around concrete tub, bathhouse

FP-52 天然方石周围的砾石，日本禅宗花园
Gravel around natural and squared stone, Japanese Zen garden

FP-53 由河岩和水泥制成的铺路石，印度尼西亚庭院
Paving stone made from river stone and cement, Indonesian courtyard

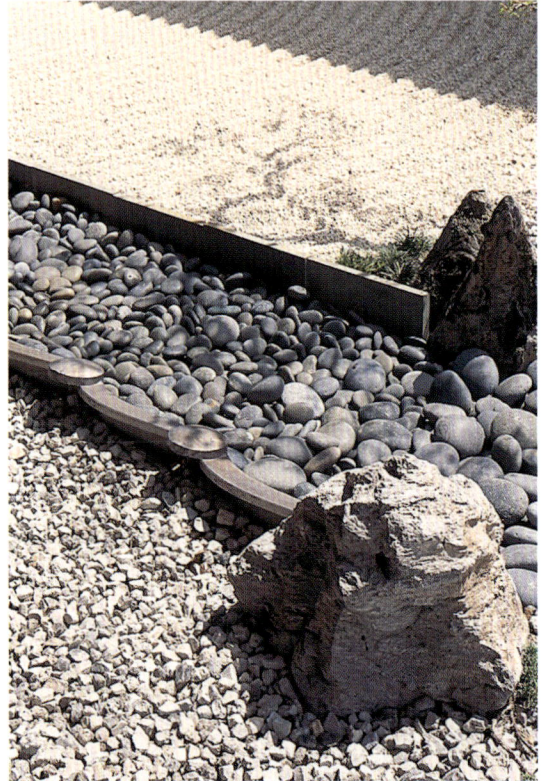

FP-54 带有海滩卵石的岩石花园细部，位于大砾石（底部）和耙平的细小砾石（顶部）之间
Detail of rock garden with beach pebbles between large gravel (bottom) and fine raked gravel (top)

石头地面与铺装

FP-55　被砾石包围的踏脚石，住宅入口通道
Steppingstones surrounded by gravel, residential entryway

摄影：Austin, Patterson, Disston

FP-56　从左到右：沥青路上的油砥石；砂浆砌合的顺砖压缝砌合圆石；纽约蓝灰
砂岩台阶，住宅入口通道
Left to right: oiled stone gravel on asphalt paving; mortared running-bond cobblestones;
New York bluestone step, residential driveway

FP-57　以苔藓和混凝土或石头为镶边的海滩卵石
Beach pebbles bordered by moss and concrete or stone

石 头 地 面 与 铺 装

地板和铺装

FP-58　带立砌收边时顺砖砌合的铺路砖
Brick pavers in running bond bordered by soldier courses intersecting pebble path; flexible paving, garden walk

FP-59　砂浆砌合铺路石的顺砖砌合变化，住宅车道
Running-bond variations in mortared brick paving, residential driveway

FP-60　从顶部到底部：平接式安装的墨西哥铺路砖，浆砌的人行道和立砌的台阶，立砌的路缘，住宅入口
Top to bottom: Mexican paving tile, straight-joint installation; mortared brick pavement and steps in running bond; soldier-course brick curb, residential entry

砖铺人行道

摄影：Austin, Patterson, Disston

FP-61 顺砖砌合的比利时式混合砖；排砖立砌的缘石和红木面的铺面板，住宅露台

Belgian-blend brick in running bond; soldier-course border and facing of mahogany decking, residential patio

FP-62 对缝砌合砖铺路和大型砂石混凝土交替排列，城市公共空间

Brick pavers in stack bond alternating with wood planks and large-aggregate concrete, urban public space

FP-63 古老人行道上的人字形图案边石和罗马砖

Stone curb and Roman brick in herringbone pattern on ancient sidewalk

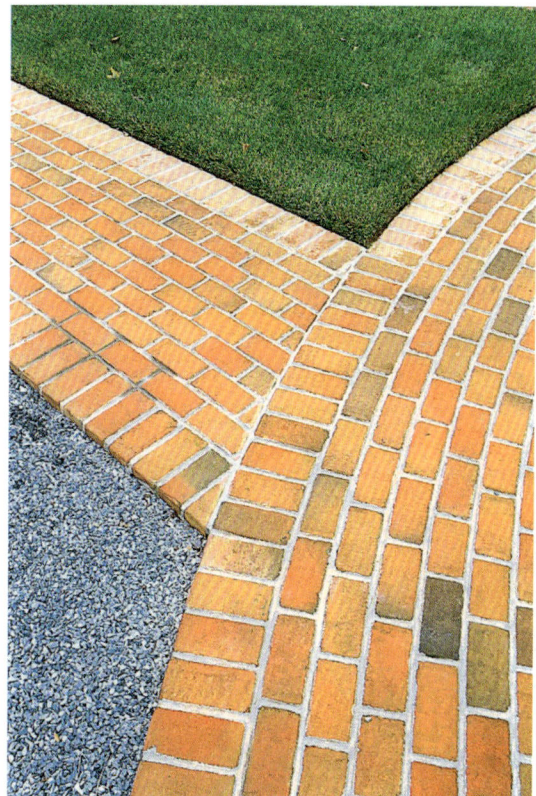

摄影：Austin, Patterson, Disston

FP-64 顺砖砌合的砂浆砌合砖，用于与卵石路和草地交叉的笔直或弯曲的小路上，带立砌收边，花园通道

Mortared brick in running bond, bordered by soldier courses in straight and curved paths intersecting with pebble path and grass, garden walk

砖铺人行道

FP-65　风化的砖铺地的俯视图，古罗马废墟
Overhead view of weathered brick paving, ancient Roman ruins

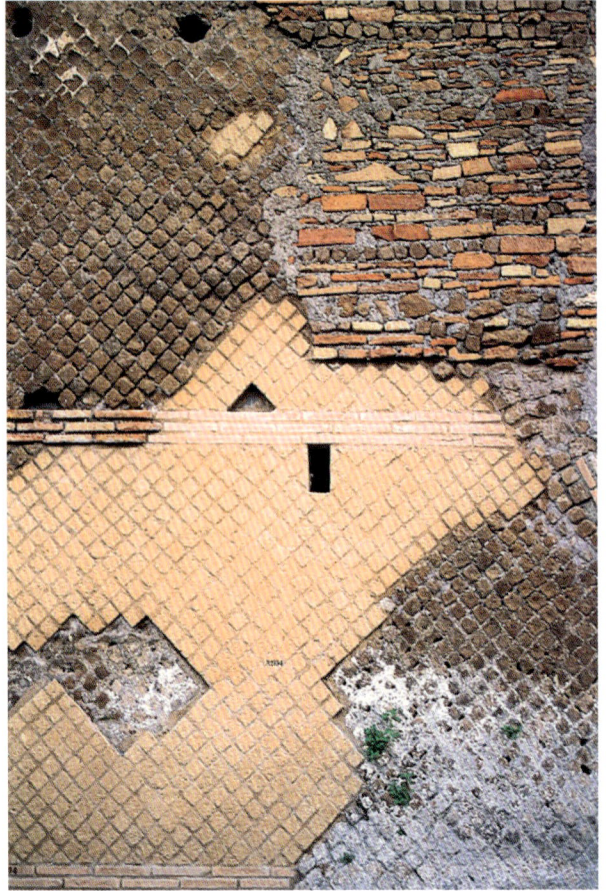

FP-66　风化砖瓦铺面的俯视图，古罗马废墟
Overhead view of weathered brick and tile paving, ancient Roman ruins

FP-67　人字形图案的铺路砖，灵活铺砌
Paving brick in herringbone pattern; flexible paving

砖铺人行道

FP-68　有石头边饰的菱形图案卵石嵌在混凝土中
Pebbles set in concrete in diaper pattern with stone borde

FP-69　深色洗石面中的浅色洗石面预浇铸铺路
Pre-cast paving blocks of light-colored exposed aggregate in a field of dark-colored exposed aggregate

FP-70　入口台阶和地面上的带有赤陶砖的洗石面，所采用的卵石为石，非水磨
Exposed aggregate with terracotta tile insets on steps and floor in entryway. Aggregate applied, not revealed by washing

摄影：Duane Langenwalter

FP-71　有瓷砖边饰的洗石面，所采用的卵石为，非水磨
Exposed aggregate with tile border. Aggregate applied, not revealed by washing

骨料与瓷砖

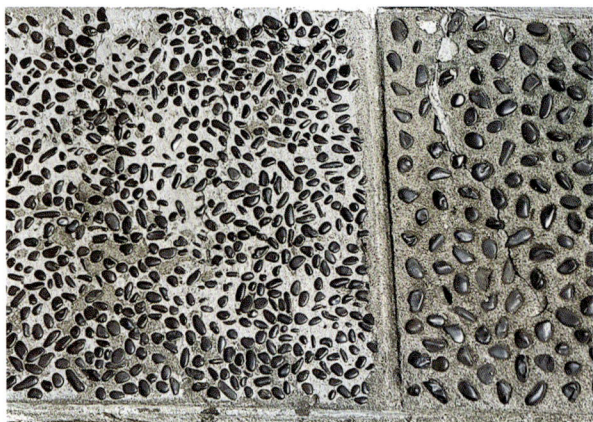

FP-72 相邻的中卵石洗石面（左）和大卵石洗石面（右），所采用的卵石为非水磨
Adjacent squares of medium (left) and large (right) exposed aggregate. Aggregate applied, not revealed by washing

FP-73 现代商业建筑庭院中被金属或混凝土格分开洗石面可能为预浇铸
Squares of exposed aggregate (possibly pre-cast) separated by metal or concrete grid in courtyard of modern commercial building

FP-74 豌豆大小的碎砾砂石（右）紧邻着形状各异随意放置的瓷砖
Pea-sized gravel aggregate (right) next to randomly shaped and placed tile

FP-75 大洗石为面的弯曲台阶，所采用的卵石为非水磨
Curved steps faced with large exposed aggregate. Aggregate applied, not revealed by washing

FP-76 豌豆大小的碎砾砂石嵌饰镶在岩石带之中，旁边的混凝土中为小裂缝光面粗石
Pea-sized gravel aggregate inset with rows of rocks and bordered by small split, polished fieldstone set in concrete

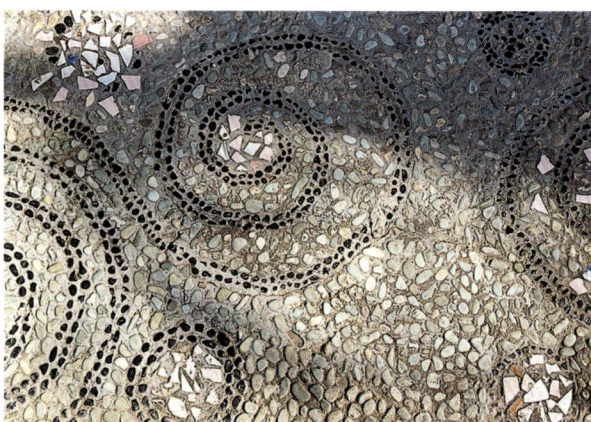

FP-77 洗石面中的彩色卵石弯曲设计
Serpentine design of polychrome pebbles in exposed aggregate

骨料与瓷砖

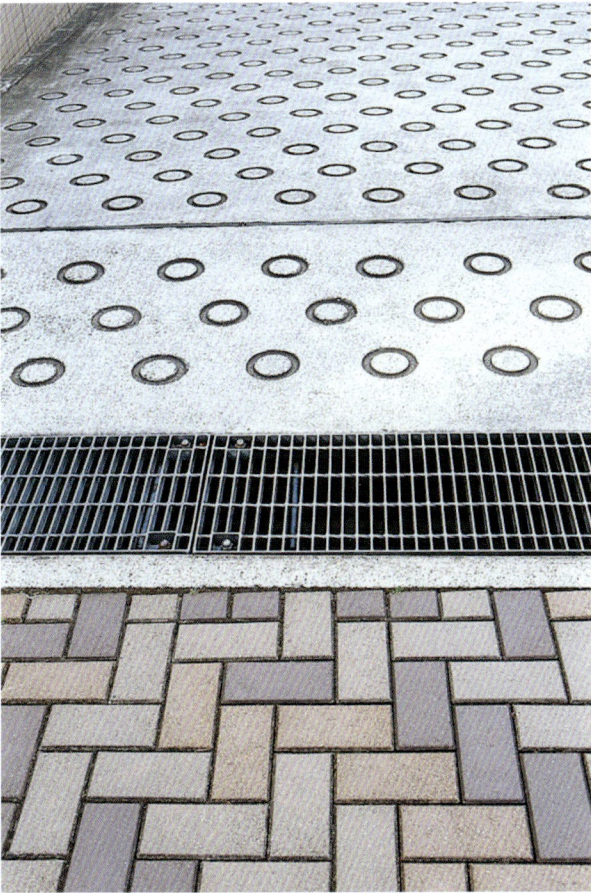

FP–78 从顶部到底部：印于混凝土中的圆形图案，金属格栅排水盖，人字形图案的预浇铸彩色混凝土铺路

Top to bottom: circle diaper pattern stamped in concrete; metal grate drain cover; pre-cast polychrome concrete pavers in herringbone pattern

FP–79 看起来类似粗面木板纹理的混凝土，很可能是现浇筑的，混凝土中的纹理是用模板材料形成的

Concrete textured to resemble rough wood planks, possibly cast-in-place, with texture produced by forming materials

FP–80 人行道上模仿石头铺路图案和脚印图案的现浇筑混凝土

Cast-in-place concrete stamped with pattern of stone paving and footprint in sidewalk

骨料与瓷砖

FP-81 彩色沥青图案，灵活铺设，城市公共空间
Polychrome asphalt pattern; flexible paving, urban public space

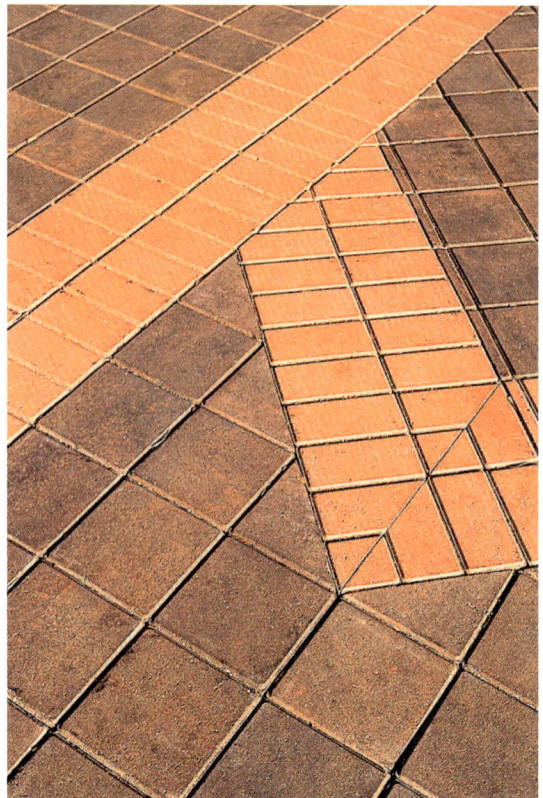

FP-82 预浇筑混凝土铺路，平接的双立砌式砖，柔性路面，城市公共空间
Pre-cast concrete pavers; double soldier course in straight-joint-course field; flexible paving, urban public space

FP-83 彩色预浇筑混凝土纹理铺路：马赛克区域有平接的边饰，城市公共空间
Polychrome pre-cast concrete textured pavers: mosaic area (top) bordered by straight-joint border, urban public space

FP-84 人字形和带状图案的彩色预浇筑混凝土铺路，城市街道
Polychrome pre-cast concrete pavers in herringbone and banded patterns, urban street

骨料与瓷砖

FP-85　有排水口的几何图案的彩色预浇筑混凝土纹理铺路材料，城市公共空间
Polychrome pre-cast concrete pavers in geometric pattern with drain, urban public space

FP-86　乱砌琢石图案的预浇筑混凝土嵌砌铺路，柔性路面（左）与陶土砖和排水盖相邻，城市公共空间
Tumbled pre-cast concrete pavers in random-ashlar pattern; flexible paving (left) bordered by terracotta tile and drain grate, urban public space

FP-87　石砖和随意形状的陶瓷镶嵌在被乱砌琢石铺路，城市公共空间
Random-shaped ceramic and stone tiles in mosaic pattern surrounded by random-ashlar pavers, urban public space

FP-88　混凝土屏挡墙被用在露台的地面上，并且在孔洞中种草
Concrete screen-wall blocks set into the ground and planted with grass in patio

FP-89　柔性路面：人字形图案的混凝土铺路（上），带有边饰带；顺砖砌合的任意大小混砌的人字形图案（下），城市公共空间
Flexible paving: concrete pavers in herringbone pattern (top), with border course; tumbled random-size concrete pavers in running bond (bottom), urban public space

FP-90　带状和星形设计的水磨石，酒店入口
Terrazzo in border and star design, hotel entryway

骨料与瓷砖

FP-91 被顺砖砌合的石板环绕装饰性上釉瓷砖，商业建筑的入口
Decorated glazed ceramic tiles surrounded by stone in running bond, commercial entry

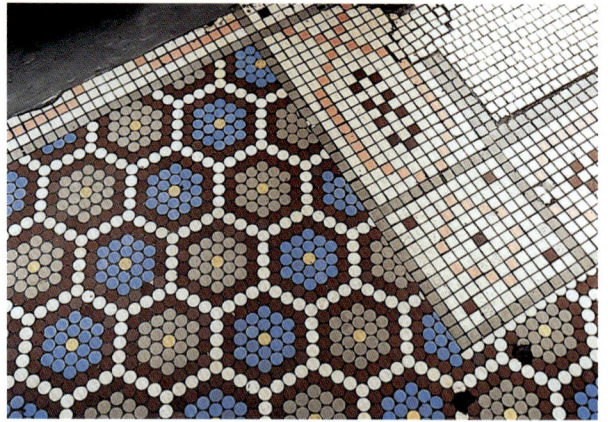

FP-92 以六边形图案镶嵌的圆瓷砖，带有方形边饰瓷砖，商业建筑的入口
Round porcelain tile mosaic in hexagon pattern with square border tiles, commercial entry

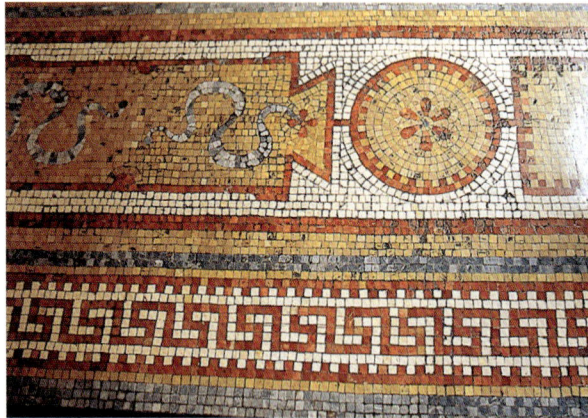

FP-93 方形石砖镶嵌在新古典风格的设计和希腊回纹饰边饰之中，商业建筑的入口
Square stone tile mosaic in neoclassical design and Greek-key border, commercial entry

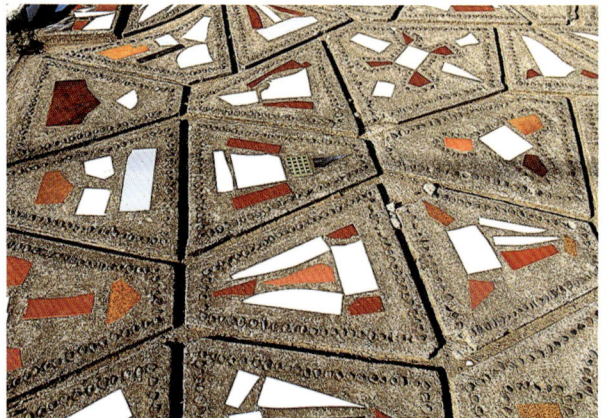

FP-94 镶嵌有瓷砖和卵石的梯形混凝土铺路，露台
Trapezoidal concrete pavers inlaid with ceramic tiles and pebbles, patio

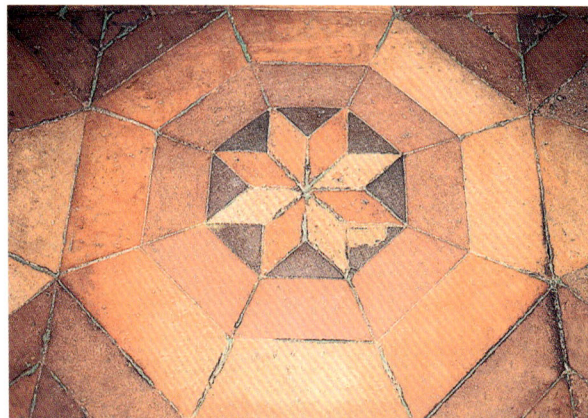

FP-95 有中心装饰的八边形边框中的铺路瓷砖
Paving tiles in octagonal border with central ornament

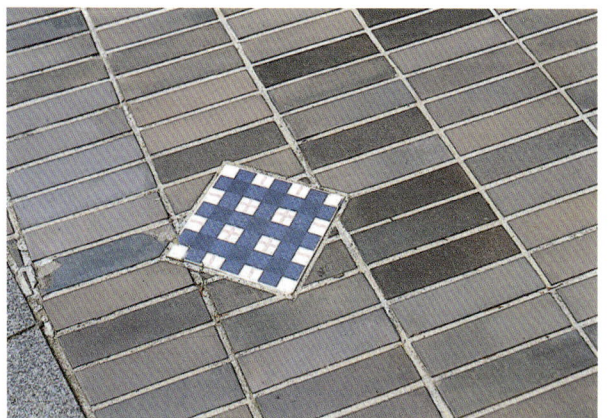

FP-96 平缝瓷砖地面上的方形装饰性上釉瓷砖，现代商业建筑的入口
Square decorated glazed tile in a field of straight-jointed ceramic tiles, modern commercial entry

FP-97 不规则的瓷砖块嵌在凸起的混凝土中，露台
Irregular pieces of ceramic tile set in raised concrete, patio

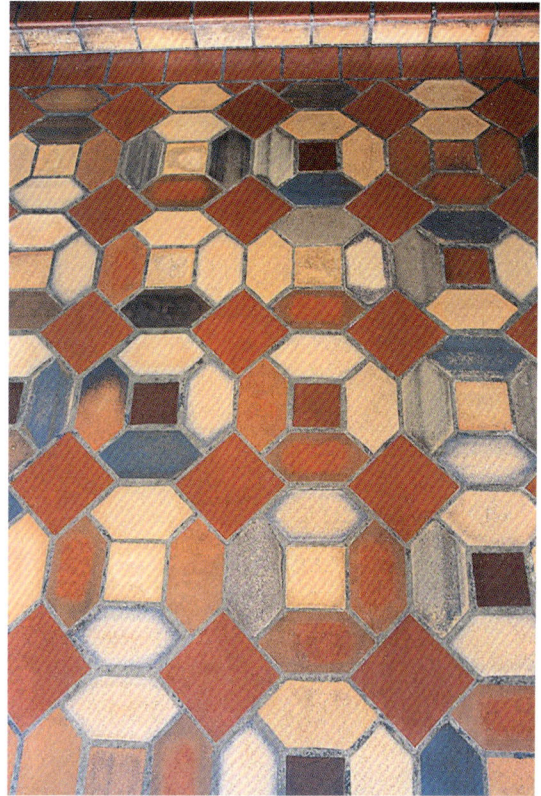

FP-98 几何图案中的彩釉陶土瓷砖
Polychrome glazed terracotta tile in geometric pattern

FP-99 平缝墨西哥式铺路瓷砖，镶嵌有装饰性的菱形上釉小瓷砖，竖缝边饰外框（右）
Mexican paving tiles, straight-jointed, and inset with small decorated glazed tiles in diaper pattern, vertical-joint border (right)

FP-100 棋盘式图案的方形预灌浆马赛克砖块，带有深色马赛克拼成的轮廓线，商业建筑入口
Squares of pre-grouted stone mosaic tiles in checker design with dark mosaic grid outline squares, commercial entry

FP-101 几何图案和乱砌图案的方形和三角形瓷砖嵌在洗石面上，入口
Square and triangular ceramic tiles set in exposed-aggregate field in geometric and random patterns, entryway

FP-102 几何图案的三角形瓷砖，城市公共空间
Triangular ceramic tiles in geometric pattern, urban public space

骨料与瓷砖

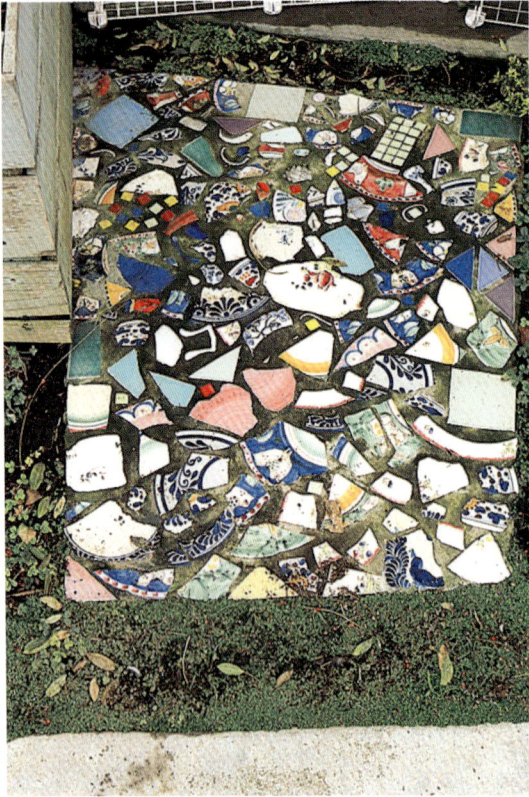

FP-103　马赛克踏脚石中用作镶嵌块的装饰性瓷盘的碎片，
住宅入口
Shards of decorated ceramic plates used as tesserae in mosaic
steppingstone, residential entryway

FP-104　马赛克格子设计中用作镶嵌块的任意形状的瓷砖片
Random-shaped tile pieces used as tesserae in mosaic grid
design set in concrete

FP-105　平缝装饰性上釉瓷砖，组成了菱形的图案和重复图案边
饰
Decorated glazed tile, straight-jointed, forming diaper patterns,
and a running-repeat border

FP-106　有字母、标志和装饰带的马赛克地面，酒店入口
Mosaic floor with lettering, logo, and decorative border, hotel
entry

骨料与瓷砖

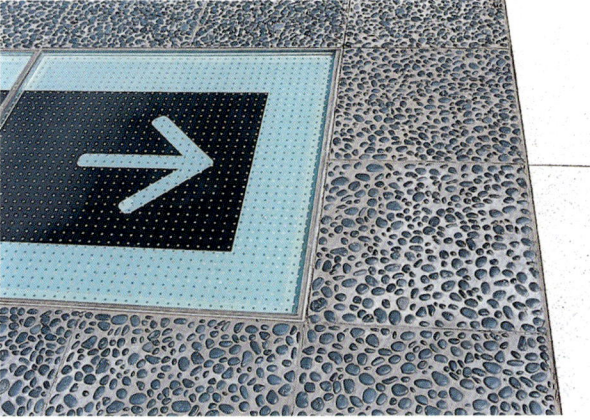

FP-107 有金属带边框的盲道中带有指示方向的箭头，周围是预浇筑的洗石面，城市公共空间
Directional arrow set in safety paving with metal strip border, surrounded by pre-cast exposed aggregate, urban public space

FP-108 混凝土中的石板上的金属角钉盲道，城市公共空间
Metal safety-paving brads on flagstone set in concrete, urban public space

FP-109 金属角钉盲道和金属长条形盲道镶嵌在花岗岩铺路中，城市公共空间
Metal safety-paving brads and strips applied to granite pavers, urban public space

FP-110 带有装饰性图案和字母的铸铁检查井井盖
Cast-iron manhole cover with decorative pattern and lettering

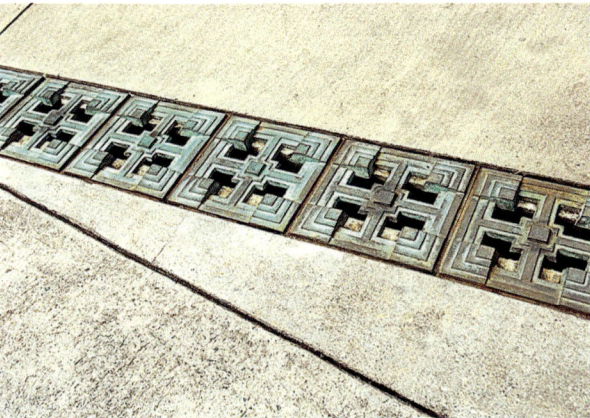

FP-111 混凝土车道上的装饰性的有铜绿的青铜排水沟盖
Ornamented and patinated bronze drain covers in concrete driveway

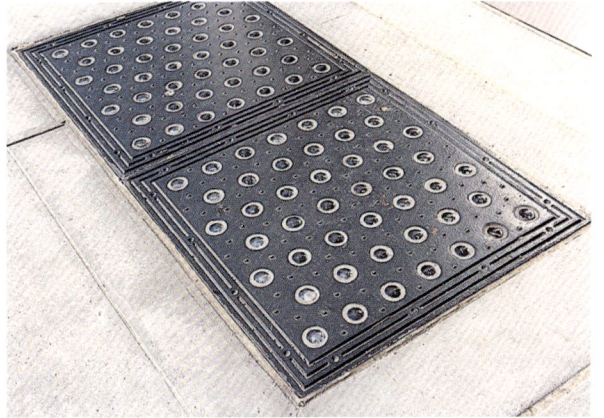

FP-112 镶嵌有彩色玻璃的钢板，人行道上的装载升降机的盖
Steel plates with colored glass insets, sidewalk loading lift cover

排水口与路线

FP-113 人行步道旁预浇筑洗石面的排水箅，城市公园
Pre-cast exposed-aggregate drain grates in gutter beside a walkway, urban park

FP-114 装饰有玫瑰花图案的铸铁检查井井盖
Cast-iron manhole cover decorated with rosettes

FP-115 生锈的铸铁排水箅
Rusting cast-iron drain grate

FP-116 人行步道上的金属拉网格栅，上面有钢质脚印图案
Expanded-metal sidewalk grate with applied steel footprints

FP-117 钢排水箅嵌在规格铺路石和有天然石头边饰的沙砾中
Steel drain grate inset into dimension paving stone and gravel with natural rock border

FP-118 装饰性铸铁路牌镶嵌在铺路材料中
Decorated cast-iron street plate inset into pavers

排水口与路线